AUGUSTUS DE MORGAN

ELEMENTS OF ALGEBRA

PRELIMINARY TO THE DIFFERENTIAL CALCULUS

AND FIT FOR THE HIGHER CLASSES OF SCHOOLS IN WHICH
THE PRINCIPLES OF ARITHMETIC ARE TAUGHT

Elibron Classics
www.elibron.com

Elibron Classics series.

ISBN 1-4021-5872-6 (paperback)
ISBN 1-4021-0202-X (hardcover)

This Elibron Classics Replica Edition is an unabridged facsimile of the edition published in 1837 by Taylor and Walton, London.

ELEMENTS OF ALGEBRA

PRELIMINARY TO THE DIFFERENTIAL CALCULUS

AND FIT FOR THE HIGHER CLASSES OF SCHOOLS IN WHICH THE PRINCIPLES OF ARITHMETIC ARE TAUGHT.

BY

AUGUSTUS DE MORGAN,

OF TRINITY COLLEGE, CAMBRIDGE,
AND PROFESSOR OF MATHEMATICS IN UNIVERSITY COLLEGE, LONDON.

SECOND EDITION.

"What a benefite that onely thyng is, to haue the witte whetted and sharpened, I neade not trauell to declare, sith all men confesse it to be as greate as maie be. Excepte any witlesse persone thinke he maie bee to wise. But he that moste feareth that, is leaste in daunger of it. Wherefore to conclude, I see moare menne to acknowledge the benefite of nomber, than I can espie willyng to studie, to attaine the benefites of it. Many praise it, but fewe dooe greatly practise it: onlesse it bee for the vulgare practice, concernyng Merchaundes trade. Wherein the desire and hope of gain, maketh many willyng to sustaine some trauell. For aide of whom, I did sette forth the firste parte of *Arithmetike*. But if thei knewe how farre this seconde parte, dooeth excell the firste parte, thei would not accoumpte any tyme loste, that were imploied in it. Yea thei would not thinke any tyme well bestowed, till thei had gotten soche habilitie by it, that it might be their aide in al other studies."—ROBERT RECORDE.

LONDON:

PRINTED FOR TAYLOR AND WALTON,
BOOKSELLERS AND PUBLISHERS TO THE UNIVERSITY OF LONDON,
28 UPPER GOWER STREET.

M.DCCC.XXXVII.

LONDON:
PRINTED BY JAMES MOYES, CASTLE STREET,
LEICESTER SQUARE.

PREFACE

TO THE

FIRST EDITION.

In the title-page I have endeavoured to make it clear that it will be impossible to teach algebra on the usual plan by means of this work. It is intended only for such students as *have that sort of knowledge of the principles of arithmetic which comes by demonstration*, and whose reasoning faculties have therefore already undergone some training.

Algebra, as an art, can be of no use to any one in the business of life; certainly not as taught in schools. I appeal to every man who has been through the school routine whether this be not the case. Taught as an art it is of little use in the higher mathematics, as those are made to feel who attempt to study the differential calculus without knowing more of its principles than is contained in books of rules.

The *science* of algebra, independently of any of its uses, has all the advantages which belong to mathematics in general as an object of study, and which it is not necessary to enumerate. Viewed either as a science of quantity, or as a language of symbols, it may be made of the greatest service to those who are sufficiently acquainted with arithmetic, and have sufficient power of comprehension, to enter fairly upon its difficulties. But if, to meet the argument that boys cannot learn algebra in its widest form, it be proposed to evade the real and efficient part of the science, whether by presenting results only in the form of rules, or by omitting and taking for granted what should be inserted and proved, for the purpose of making it appear that something *called* algebra has been learned: I reply, that it is by no means necessary, except for show, that the word algebra should find a place in the list of studies of a school; that, after all, the only question is, whether what is taught under that name be worth the learning; and that if real *algebra*, such as will be at once an exercise of reasoning, and a useful preliminary to subsequent studies, be too difficult, it must be deferred. Of this I am quite sure, that the student who has no more knowledge of arithmetic—that is, of the reasoning on which arithmetical notation and processes are built—than usually falls to the lot of those who begin algebra at school—that is, I believe, begin to add *positive* and *negative* quantities together,—will sooner find his way barefoot to Jerusalem than understand the greater part of this work. And I may say the same of every work on algebra, containing reasoning and not rules, which I have ever seen; provided it contained any of the branches of the subject which are of most usual application in the higher parts of mathematics.

The special object to which this work is devoted is the developement of such parts of algebra as are absolutely requisite for the study of the differential calculus, the most important of all its applications. The former science is now so extensive, that some particular line must be marked out by every writer of a small treatise. The very great difficulty of the differential calculus has always been a subject of complaint; and it has frequently been observed that no one knows exactly what he is doing in that science until he has made considerable progress in the mechanism of its operations. I have long believed the reason of this to be that the fundamental notions of the differential calculus are conventionally, and with difficulty, excluded from algebra, in which I think they ought to occupy an early and prominent place. I have, therefore, without any attention to the agreement by which the theory of limits is never suffered to make its appearance in form until the commencement of the differential calculus, introduced limits throughout my work: and I can certainly assure the student, that, though I have perhaps thereby increased the difficulty of the subject, the additional quantity of thought and trouble is but a small dividend upon that which he would afterwards have had to encounter, if he had been permitted to defer the considerations alluded to till a later period of his mathematical course. On those who offer theoretical objections to the introduction of limits in a work on algebra lies the *onus* of shewing that they are not already introduced, even in arithmetic. What is $\sqrt{2}$, supposing geometry and limits both excluded?

I have been sparing of examples for practice in the earlier part of the work, and this because I have always found that manufactured instances do not resemble the combinations which actually occur. They are but a sort of parade exercise, which cannot be made to include the means of meeting the thousand contingencies of actual service. The only method of furnishing useful cases is to take some inverse process, and the verification of literal equations (as in the seventh and following pages of this work) is the most obvious. With these the student can furnish himself at pleasure, the test of correctness being the ultimate agreement of the two sides, after the value of the unknown quantity has been substituted.

The only remaining caution which he will need is, not to proceed too quickly, especially in the earlier part of the work. He must remember that he is engaged upon a very difficult subject, and that if he does not find it so, it is, most probably, because he does not understand what he is about. Wherever an instance or a process occurs, he should take others as like them as he can, and assure himself, by the reasoning in the work, that he has obtained a true result.

It was at first my intention to write a second volume on the higher parts of the subject. But, considering that of these there are several distinct branches, it has appeared to me that the convenience of different classes of students will be consulted by publishing them in several distinct tracts, which may afterwards be bound together, if desired.

AUGUSTUS DE MORGAN.

London,
July 31*st*, 1835.

*** This Second Edition differs from the first only in verbal amendments. Since the publication of the first edition, I have carried on the consideration of the principles of algebra in two other works. In the first, or *Elements of Trigonometry*, besides the consideration of periodic magnitudes in general, I have (Chapter IV.) given a view of that extension of the meaning of symbols which must accompany the complete explanation of the negative square root. In the second, or *Connexion of Number and Magnitude*, which is an attempt to explain the Fifth Book of Euclid, I have entered upon what is in reality the most difficult part of the application of arithmetic to geometry. Both of the preceding works may be made supplementary to the present one, though the latter is altogether independent of it.

University College, London,
October 16, 1837.

CONTENTS.

CHAPTER VIII.

CHAPTER IX.

CHAPTER X.

CHAPTER XI.

CHAPTER XII.

CHAPTER XIII.

ELEMENTS OF ALGEBRA.

INTRODUCTION.

It is taken for granted that the student who attempts to read this work has a good knowledge of arithmetic, particularly of common and decimal fractions. Whoever does not know so much had better begin by acquiring it, as the *shortest road* to algebra.*

In arithmetic, we use *symbols* of *number*. A symbol is any sign for a quantity which is not the quantity itself. If a man counted his sheep by pebbles, the pebbles would be symbols of the sheep. Our symbols are marks upon paper, of which the meaning of every one is determined so soon as the meaning of 1 is determined. If we are speaking of length, we choose a certain length, any we please, and call it 1. It may have any other name in common life, for instance, a foot or a mile, but in arithmetic, when we are numbering by means of it, it is 1. We now introduce the sign $+$, and agree that when we write $+$ between two symbols of quantity, it shall be the symbol for the quantity made by putting these two quantities together. Thus, if

1 stand for the length ————

$1 + 1$ stands for the length ————————

$1 + 1$ is abbreviated into 2, a new and *arbitrary* † symbol. Similarly $2 + 1$ is abbreviated into 3, $3 + 1$ into 4, and so on.

* The references in this work are to the *articles* (not pages) of my Treatise on *Arithmetic*, and serve either for the second or third editions.

† *Arbitrary*, that is, any other would have done as well. It is 2 that stands for $1 + 1$, and not $<$, 3, ∞, or any thing else, because certain Hindoos chose that it should be so. See *Penny Cyclopædia*, Art. Arithmetic.

When 1, 2, 3, &c., mean 1 mile, 2 miles, 3 miles, &c., or 1 pint, 2 pints, 3 pints, &c., these are called *concrete* numbers. But when we shake off all idea of 1, 2, &c., meaning *one, two*, &c., of any thing in particular, as when we say, "six and four make ten," then the numbers are called *abstract* numbers. To the latter the learner is first introduced, in regular treatises on arithmetic, and does not always learn to distinguish rightly between the two. How many of the operations of arithmetic can be performed with concrete numbers, and without speaking of more than one sort of 1? *Only addition and subtraction.* Miles can be added to miles, or taken from miles. Multiplication involves a new sort of 1, 2, 3, &c., standing for *repetitions* or *times*, as they are called. Take 6 miles 5 times. Here are two kinds of units, 1 mile and 1 time. In multiplication, one of the units must be a number of repetitions or times, and to talk of multiplying 6 feet by 3 *feet*, would be absurd.* What notion can be formed of 6 feet taken "3 feet" times?

But in solving the following question, "If 1 yard cost 5 shillings, how much will 12 yards cost?" do we not multiply the twelve *yards* by the five *shillings*? Certainly not—the process we go through is the following: Since each yard costs five shillings, the buyer must put down 5 shillings as often (as many *times*) as the seller uses a one-yard measure; that is, 5 shillings is taken 12 *times*.

In division, we must have the idea either of repetition or of *partition*, that is, of cutting a quantity into a number of equal parts. "Divide 18 miles by 3 *miles*," means, find out how many *times* 3 miles must be repeated to give 18 miles: but "Divide 18 miles by 3," means, cut 18 miles into three equal parts, and find how many miles are in each part.

18 miles divided by 3 *miles* gives 6; meaning, that 3 *miles* must be repeated six *times* to give 18 *miles*.

18 *miles* divided by 3 gives 6 *miles*; meaning, that if 18 miles be cut into three *equal parts*, each part is 6 *miles*. The answer in abstract numbers is the same in both cases; 18 divided by 3 gives 6.

But now we ask, How many times does 12 feet contain 8 feet?

* In old books the following is sometimes found. "What is £99. 19*s*. 11$\frac{3}{4}$*d*. multiplied by £99. 19*s*. 11$\frac{3}{4}$*d*.?" The only intelligible meaning of this is as follows: If a stock of money is to be increased at the rate of £99. 19*s*. 11$\frac{3}{4}$*d*. for every £1 in it, how much will that be when the stock itself is £99. 19*s*. 11$\frac{3}{4}$*d*.? Let the student answer this.

The answer is, more than once and less than twice; which is not complete, because we have not an adequate idea of parts of times, that is, parts of repetitions. In talking of *times*, we use a figure of speech which we may liken to a machine which works by starts, each start doing, say eight feet of work, and which is so contrived that nothing less than a whole start can be got from it; either 8 feet or nothing. It is plain that such a machine cannot execute 12 feet of work, or any thing between 8 feet and 16 feet. But, let us now suppose the machine to be made to work regularly, at 8 feet a minute. Let us still continue to call 8 feet a start, then 12 feet must be called a start and a half. In the same way we say that 12 feet contains 8 feet a *time* and a *half*, the notion of half a time being equivalent to that of repeating not the whole 8, but its half.

When we speak of dividing one fraction by another in arithmetic, this is what is meant; for instance, $\frac{2}{3}$ divided by $\frac{5}{7}$ gives $\frac{14}{15}$; or $\frac{2}{3}$ contains $\frac{5}{7}$, $\frac{14}{15}$ of a time. Let the learner study the following propositions.

If $\frac{5}{7}$ of £1 were gained in a day, then $\frac{2}{3}$ of a pound would be gained in $\frac{14}{15}$ of a day.

If the signification of 1 be changed, so that what was $\frac{5}{7}$ is now 1, then what was $\frac{2}{3}$ is now $\frac{14}{15}$.

If $\frac{5}{7}$ of the line A be the line B, then $\frac{2}{3}$ of the line A is $\frac{14}{15}$ of the line B.

The want of a proper comprehension of such questions as the preceding is a great source of difficulty to most beginners in algebra. If the preceding pages be not readily understood, it is a sign that the reader is not sufficiently acquainted with arithmetic for his purpose in reading this work.

The symbols of ARITHMETIC have a *determinate connexion;* for instance, 4 is always $2+2$ whatever the things mentioned may be, miles, feet, acres, &c. &c. In ALGEBRA, we take symbols for numbers which have no determinate connexion. As in arithmetic we draw conclusions about 1, 2, 3, &c., which are equally true of 1 foot, 2 feet, &c., 1 minute, 2 minutes, &c.; so in algebra we reason upon numbers in general, and draw conclusions which are equally true of all numbers. This, at least, is one great branch of algebra, and exhibits it in a view most proper for a beginner.

But this is a definition in a few words, and can only be understood by those who have already studied the subject. No science can be defined in a few words to one who is ignorant of it. We shall begin by giving an instance of a general property of numbers and fractions.

Take (8) units,* and the fraction which, taken (8) times, gives a unit, that is the (8)th part of a unit. Add 1 to both; this gives 9 and $1\frac{1}{8}$. The first contains the second (8) times. Take $\left(\frac{2}{3}\right)$ of a unit, and the fraction which taken $\left(\frac{2}{3}\right)$ of a time gives a unit, or $1\frac{1}{2}$. Add 1 to both, this gives $1\frac{2}{3}$ and $2\frac{1}{2}$. The first contains the second $\left(\frac{2}{3}\right)$ of a time. Try the following, in which it will be found that the blanks may be filled up with any one number or fraction at pleasure.

Take ( ) units, and the number or fraction which repeated ( ) times gives a unit. Add 1 to both; then the first result will contain the second ( ) times, or parts of times.

The following are instances which should be tried.

( )	Number or fraction which repeated ( ) times gives a unit.	1 added to the first.	1 added to the second.	Number of times which the third contains the fourth.
7	$\frac{1}{7}$	8	$1\frac{1}{7}$	7
$\frac{1}{3}$	3	$1\frac{1}{3}$	4	$\frac{1}{3}$
$2\frac{1}{4}$	$\frac{4}{9}$	$3\frac{1}{4}$	$1\frac{4}{9}$	$2\frac{1}{4}$
$\frac{1}{20}$	20	$1\frac{1}{20}$	21	$\frac{1}{20}$
1	1	2	2	1

The connexion between the first and second columns is this, that the number or fraction in the second is 1 divided by the number or

* By putting 8 in brackets, we wish to call attention to the circumstance of the numbers in the different brackets being the same.

fraction in the first; or the number of times or parts of times which 1 contains the first. That is, if we call the number in the first column

"the number,"

then the number in the second column is

1 divided by "the number."

And the coincidence of the first and fifth columns (which constitutes the thing we notice) may be thus expressed:

Let one more than "the number" be divided by "1 more than the times which 1 contains the number," and the result must be "the number."

The above must be still further abbreviated for convenience. As "the number" means any number we please, and as "any number we please" will be better expressed by any short symbol which we may choose to make use of, let one of the letters of the alphabet be employed, say a. Let the addition of a number be denoted by + as before, and let the division of a number by a number be denoted, as in arithmetic, by writing the divisor under the dividend with a line between them. Let = be the sign that what goes before is the same number as what comes after. Then the preceding property of numbers is thus expressed:

$$\frac{a+1}{\frac{1}{a}+1} = a$$

We shall now proceed to lay down the definitions of the first part of the science.

I. Algebra is the European corruption of an Arabic phrase, which may be thus written, *al jebr e al mokabalah*, meaning *restoration and reduction.* The earliest work on the subject is that of Diophantus, a Greek of Alexandria, who lived between A.D. 100 and A.D. 400; but when, cannot be well settled, nor whether he invented the science himself, or borrowed it from some Eastern work. It was brought among the Mahometans by Mohammed ben Musa (Mahomet, the son of Moses) between A.D. 800 and A.D. 850, and was certainly derived by him from the Hindoos. The earliest work which has yet been found among the latter nation, is called the *Vija Ganita*, written in the Sanscrit language, about A.D. 1150. It was introduced into Italy, from the Arabic work of Mohammed, just mentioned, about the beginning of the thirteenth* century, by Leonardo Bonacci, called Leonard of Pisa: and into England by a physician, named Robert Recorde, in a book called the *Whetstone of Witte*, published in the reign of Queen Mary, in 1557. From this work the motto in the title-page is taken.

II. A letter denotes a number, which may be, according to circumstances, as will hereafter appear, either any number we please; or some particular number which is not known, and which, therefore, has a sign to represent it till it is known; or some number or fraction which is known, and is so often used that it becomes worth while to have an abbreviation for it. Thus the Greek letters π and ε always stand for certain results, which cannot be exactly represented, but which are nearly 3·1415927 and 2·7182818.

III. The alphabets used are 1. The *Italic* small letters; 2. The Roman capitals; 3. The Greek small letters; 4. The Roman small letters; 5. The Greek capitals. They are here placed in the order of their importance on the subject; and as many may wish to learn algebra, who do not know Greek, the Greek alphabet is here given, with the pronunciation of the letters.

Α	α	alpha	Ν	ν	nu
Β	β ϐ	bēta	Ξ	ξ	xi
Γ	γ	gamma	Ο	ο	ŏmīcron
Δ	δ	delta	Π	π ϖ	pi
Ε	ε	epsīlon	Ρ	ϱ	ro
Ζ	ζ ζ	zēta	Σ	σ ς	sigma
Η	η	hēta	Τ	τ ſ	tau
Θ	θ	thēta	Υ	υ	ūpsīlon
Ι	ι	iōta	Φ	φ	phi
Κ	κ	kappa	Χ	χ	chi
Λ	λ	lambda	Ψ	ψ	psi
Μ	μ	mu	Ω	ω	ōmĕga

IV. Under the word number is always included *whole numbers*

* The young reader may need to be told that the thirteenth century does not mean A.D. 1300 and upwards, but A.D. 1200 and upwards. The *first* century is from the beginning of A.D. 1 to the end of A.D. 100: the second from the beginning of A.D. 101 to the end of A.D. 200, and so on.

and *fractions*. Thus, $2\frac{1}{2}$ is not called a number in common language, but in algebra it *is* called a number; or, if it be necessary to distinguish it from 2, 3, 4, &c., it is called a *fractional* number, while the latter are called *whole* numbers.

Under the word number the symbol 0 may be frequently included, which means *nothing*, or the absence of all magnitude or quantity. If we ask, what number remains when b is taken from a, and if we afterwards find out that b and a must stand for the same number, then the answer is, "no number whatever remains, or there is nothing left." If we say that 0 remains, we make the symbol 0, an answer to a question beginning, "What number, &c.?" which is equivalent to including 0 under the general word *number*.

V. The sign $+$ is read *plus* (Latin for *more*), means in correct English* "increased by," and signifies that the second-mentioned number is to be added to the first. Thus $a+b$ is read *a plus b*, and means a increased by b, or the number which is made by adding b to a.

$+a$ by itself can only mean a added to nothing, or $0+a$, which is a itself.

VI. The sign — is read *minus* (Latin for *less*), means "diminished by," and implies that the second number is to be taken away from the first. Thus $a-b$ is read *a minus b*, means a diminished by b, and signifies that b is to be taken away from a.

When a is less than b, the preceding stands for nothing at all, but a direction to do what cannot be done. Thus $3-6$ is impossible. We shall hereafter have to investigate what such an answer means, that is, the problem which gave it being of course impossible, what *sort* of absurdity gives rise to the impossibility.

VII. The sign $\times$ is rendered by the English word *into*, and

* Those who have attempted to substitute a short English term have never been able to find one which does not shock the ear. We have seen "*a* add *b*," and "*a* more *b*;" which should be, "To *a* add *b*," or "*b* more than *a*." Neither of the last would be convenient, and "*a* increased by *b*" is too long. The same remarks may be made on "*a* less *b*," and "*a* take away *b*," for "*a* diminished by *b*:" while "*a* save *b*," used by our oldest writers, is now too uncommon. There is no language in which the simplest relations are expressed by the simplest terms.

means "is the multiplier of," meaning that the second number is to be taken as many times, or parts of times, as there are units, or parts of units, in the first. Thus $a \times b$ is read *a into b*, and means b taken a times. Thus, $1\frac{1}{2} \times 6$ is 6 taken a time and a half, or 9. The expression ab is called the product of a and b; the two latter are called *factors* of the product, and *coefficients* of each other. $0 \times a$ and $a \times 0$ must both mean 0; for a taken no time, nor part of a time whatsoever, cannot give any quantity, and nothing, however often repeated, yields nothing. As this is connected with a mistake always made by beginners, we shall (to impress the results on his memory) give two problems* the answer to which is self-evident, and put the answers in an algebraical form; desiring the learner, of course, to remember that no new information is gained, but only an opportunity of making certain results occupy a space proportional to their importance.

There is a number of boxes, none of which contain any thing. How much do all together contain?

If a be the number of boxes, then 0 repeated a times, or $a \times 0$ is 0.

There is a box full of gold, of which no part whatsoever belongs to A. How much belongs to A?

If p be the number of pounds of gold in the box, then A's part is $0 \times p$, or 0.

The reason of the mistake is, that the beginner retains the notion that "not multiplied" implies "not diminished," and "not changed at all." But *multiplication* in the arithmetic of fractions, and in algebra, means the taking a number of times, or *parts of times*. A number *not multiplied at all* yields no number, for no part of it is taken; a number which remains the same is multiplied *by* 1, *or taken once*. Thus a is $1 \times a$.

When letters are employed, or numbers and letters together, the sign $\times$ is dropped. Thus ab means b taken a times; $3a$ means a taken 3 times. It is only when two numbers are employed that $\times$ becomes necessary. Thus 3×6 must not be written 36, which already stands, not for $6 + 6 + 6$, but for $3 \times \text{ten} + 6$. It is more common to place points between numbers (thus, 3.6) to signify multiplication. But

* Problem, any question in which something is required to be found.

as this may be confounded with $3+\frac{6}{10}$, the student should always write the decimal point higher up, thus 3·6.

VIII. The bar between the numerator and denominator of a fraction is read "*by*," and this is the word for division. Thus $\frac{a}{b}$ is read "*a by b*," and means that it is to be found what number of times and parts of times *a* contains *b*. Thus "$\frac{3}{2}$, or 3 by 2, is $1\frac{1}{2}$," means that 3 contains 2 a time, and half a time. That a *half* is written $\frac{1}{2}$, or "1 by 2," is consistent, because it is the part of a time which 1 contains 2. The division of *a* by *b* will sometimes be denoted by $a \div b$.

$\frac{a}{0}$ (which beginners usually confound with *a*) has no meaning. How many times does six contain nothing? The answer is, that the question is not rational. *a* is $\frac{a}{1}$, for by *a* we mean a number of units and parts of units, so that *a* itself is the answer to "how many times, and parts of times, does *a* contain 1?"

IX. The following are then synonymes for *a*, which the student should repeat till he is very familiar with them:

$$\begin{array}{lllll} a+0 & a\times 1 & a\times 1\times 1 & \frac{a}{1} & \&c. \\ 0+a & & & & \\ a-0 & 1\times a & 1\times a\times 1 & \frac{a\times 1}{1} & \&c. \end{array}$$

X. The following abbreviations are used for the connecting words *equals* and *therefore*. By $a=b$ we mean that *a* and *b* are the same numbers: it is read *a* equals *b*. By $\therefore$ we mean *therefore*, or *then*, or *consequently*. Thus, $a=b$ and $b=c \therefore a=c$ is read *a* equals *b*, and *b* equals *c*, therefore *a* equals *c*.

XI. Every collection of algebraical symbols is called an *expression*, and when two expressions are connected by the sign $=$, the whole is called an *equation*.

An *identical* equation is one in which the two sides must be always equal, whatever numbers the letters stand for, such as the equation already described (page iv),

$$\frac{a+1}{\frac{1}{a}+1}=a$$

which cannot be untrue for any value of a; or is true, whatever number we may suppose a to stand for.

The following are very obvious identical equations:

$$a+b=b+a \qquad a+1+1=a+3-1$$

$$a+a=2a \qquad a+a+a=3a$$

$$\frac{1}{2}a+\frac{1}{2}a=a \qquad \frac{1}{3}a+\frac{1}{3}a+\frac{1}{3}a=a$$

$$2a+3a-a+4a=8a+6a+5a-11a$$

The following are not obvious, but will be found to be true in every case in which they are tried.

$$\frac{1}{1+a}+\frac{1}{1+2a}=\frac{2+3a}{1+3a+2aa}$$

$$a-\frac{a}{1+x}=\frac{ax}{1+x}$$

An *equation of condition* is one which is not always true for every value of the letters, but only for a certain number of values. Thus

$$b+1=7 \quad \text{and} \quad a-3=12$$

cannot be true unless b is 6 and a is 15.

Again, $a=b+c$ cannot be true for every value of a, b, and c, but requires or lays down the condition that a must be the sum of b and c. When a number may be written for a letter in an equation, and it remains true, that number is said to *satisfy* the equation. Thus, $a-3=12$ is satisfied by $a=15$.

XII. When an algebraical expression is enclosed in brackets, it signifies that the whole result of that expression stands in the same relation to surrounding symbols as if it were one letter only. Thus,

$$a-(b-c)$$

means that from a we are to take, not b, or c, but $b-c$, or what is left after taking c from b. It is not, therefore, the same as $a-b-c$.

EXAMPLE. What is $a-(b-\{c-d\})$ when $a=20$, $b=12$, $c=10$, and $d=3$. Here $c-d$ is $10-3$, or 7; $b-(c-d)$ is $12-7$, or 5; $a-(b-\{c-d\})$ is $20-5$, or 15.

Also, $(a+b)(c+d)$ means that $c+d$ is to be taken $a+b$ times, and $p(q+r)$ that $q+r$ is to be taken p times.

These are the first outlines of algebraical notation. Others will develope themselves as the work proceeds. We shall now lay down some very evident characters of the four principal processes.

1. Additions may be made in any order, without affecting the result. It is evident that all the following six expressions are the same:

$$1+2+3 \qquad 1+3+2$$
$$2+3+1 \qquad 3+2+1$$
$$3+1+2 \qquad 2+1+3$$

Also $a+b+c+d = b+c+d+a = b+a+d+c$, &c.

2. The order of additions and subtractions may be changed in any way which will not produce an attempt to subtract the greater from the less. Thus, if a man lose £20 and gain £50, he has the same as if he first gained £50 and then lost £20, with this exception, that if his property be less than £20, he may do the second but cannot do* the first. Thus $10-20+50$ is impossible; but $10+50-20$ is possible. Also $8-6+10-11$ admits of the following forms:

$$8-6+10-11 \qquad 8+10-11-6 \qquad 10+8-11-6$$
$$8+10-6-11 \qquad 10+8-6-11 \qquad 10-6+8-11;$$

but does not admit of the following:

$$10-11+8-6 \qquad -11+8+10-6 \quad \text{\&c.}$$
$$10-11-6+8 \qquad -6-11+8+10 \quad \text{\&c.}$$

QUESTION. What are the conditions necessary to the possibility of $a-b+c-d$? ANSWER. That a be greater than (or at least not less than) b, and that $a-b+c$ be not less than d. We shall hereafter return to this point.

3. Multiplications and divisions may be performed in any order. For instance, abc means that c is to be taken b times and the result a times: the rules of arithmetic shew that this is the same as bac; that is, as c taken a times, and the result taken b times; and this whether the numbers be whole or fractional. The student is supposed to have demonstrated these rules before he commences algebra.

It is also shewn in arithmetic that division and multiplication

* According to the language of common life a man may lose more than he has, that is, lose all that he has and incur a debt besides. But this is always on the tacit supposition that not only what he has, but what he may get, is liable for his debts. No such supposition exists in arithmetic until the meaning of words is altered; 10—20 is, in every *arithmetical* interpretation, impossible.

may be changed in the order of operation; that is, a divided by b and the quotient multiplied by c, is the same as a multiplied by c and the product divided by b. Or

$$c \times \frac{a}{b} = \frac{ca}{b}$$

In fact, owing to the extension of the meaning of words by which multiplication includes taking *parts of times* (page iii), every multiplication is a division, and every division a multiplication. For instance, to divide a by six is to take the sixth part of a, or to take a one sixth of a time, or to multiply a by $\frac{1}{6}$. Similarly, to divide a by $\frac{1}{4}$ is to ask how many fourths of a unit a contains; the answer to which is, *four* times as many times as a contains units, which multiplies a by 4.

Again, to divide 10 by $\frac{2}{3}$ is to multiply 10 by $\frac{3}{2}$. To divide 10 by $\frac{2}{3}$ is to find how often 10 contains two-thirds of a unit. Now a unit is made up of $\frac{2}{3}$ and $\frac{1}{3}$, the second of which is half the first; that 1 contains $\frac{2}{3}$, a time and a half. Therefore, 10 contains two-thirds ten times and ten halves of times, or 15 times. That is, 10 divided by $\frac{2}{3}$ is 10 taken once and a half or 10 multiplied by $1\frac{1}{2}$, that is, by $\frac{3}{2}$.

Similarly, to divide by $\frac{7}{2}$ is to multiply by $\frac{2}{7}$. How often does 10 contain the half of 7 units? Twice as often as it contains 7 units; that is, twice $\frac{10}{7}$ of a time or $\frac{20}{7}$ of a time. But $\frac{20}{7}$ is $\frac{2}{7}$ of 10, or 10 taken $\frac{2}{7}$ of a time.

The learner should practice many different cases of the following general assertions.

a multiplied into $\frac{p}{q}$ is a divided by $\frac{q}{p}$

a divided by $\frac{p}{q}$ is a multiplied into $\frac{q}{p}$

or $$\frac{p}{q} \times a = \frac{a}{\frac{q}{p}} \qquad \frac{a}{\frac{p}{q}} = \frac{q}{p} \times a$$

The first operation to which the beginner must be accustomed is the conversion of algebraical expressions into numbers, upon different suppositions as to the value of the letters. For instance, what is

$$\frac{a+b}{a-b} \text{ when } a = \frac{1}{2} \text{ and } b = \frac{2}{5}$$

$$a+b = \frac{1}{2} + \frac{2}{5} = \frac{9}{10} \qquad a-b = \frac{1}{2} - \frac{2}{5} = \frac{1}{10}$$

$$\frac{a+b}{a-b} = \frac{\frac{9}{10}}{\frac{1}{10}} = 9$$

Is $\dfrac{1+aa}{1+a} = \dfrac{aaa}{2+\frac{1}{2}a}$ true when $a = 1\frac{1}{2}$?

$$aa = \frac{3}{2} \times \frac{3}{2} = \frac{9}{4} \qquad 1+aa = 1+\frac{9}{4} = \frac{13}{4}$$

$$1+a = 1+\frac{3}{2} = \frac{5}{2} \qquad (1+aa) \div (1+a) = \frac{13}{4} \div \frac{5}{2} = \frac{13}{10}$$

$$aaa = \frac{3}{2} \times \frac{3}{2} \times \frac{3}{2} = \frac{27}{8} \qquad \frac{1}{2}a = \frac{3}{4} \qquad 2+\frac{1}{2}a = \frac{11}{4}$$

$$aaa \div \left(2+\frac{1}{2}a\right) = \frac{27}{8} \div \frac{11}{4} = \frac{27}{22}$$

But $(1+aa) \div (1+a) = \dfrac{13}{10}$

therefore the above equation is not true in this case.

What is $aa + b(a+b)$ when $a = 4 \quad b = 3$?

$$a+b = 7 \qquad b(a+b) = 3 \times 7 = 21 \qquad aa = 16$$

$$aa + b(a+b) = 16 + 21 = 37$$

But the most instructive exercise is the verification of equations which are asserted to be identical. For instance,

$$\frac{aa-bb}{a-b} = \frac{aaa+bbb}{aa+bb-ab}$$

Let $a = 4 \quad b = 2$, then $aa = 16 \quad bb = 4$.

$$\frac{aa-bb}{a-b} = \frac{16-4}{2} = (6) \qquad aaa = 64 \quad bbb = 8$$

$$\frac{aaa+bbb}{aa+bb-ab} = \frac{64+8}{16+4-8} = \frac{72}{12} = (6)$$

Let $a = \frac{2}{3} \quad b = \frac{1}{2}$.

$$\frac{aa-bb}{a-b} = \frac{\frac{4}{9}-\frac{1}{4}}{\frac{2}{3}-\frac{1}{2}} = \frac{\frac{7}{36}}{\frac{1}{6}} = \left(\frac{7}{6}\right)$$

$$\frac{aaa+bbb}{aa+bb-ab} = \frac{\frac{8}{27}+\frac{1}{8}}{\frac{4}{9}+\frac{1}{4}-\frac{1}{3}} = \frac{\frac{91}{216}}{\frac{13}{36}} = \left(\frac{7}{6}\right)$$

The following is a list of identical equations, which the learner should verify; first, by giving whole values to the letters, next, by giving fractional values. It is needless to give instances of each, because the test of correctness is in the two sides shewing the *same* value, no matter what. The only restriction upon the values of the letters is that no values must be assumed which will produce in the equation an expression for the subtraction of the greater from the less, or which will produce 0 in the denominator of a fraction (pages vii, ix).

$$(a+x+y)(a+x-y) = aa+2ax+xx-yy$$

$$(a+b)(a+b) = aa+2ab+bb$$

$$(a-b)(a-b) = aa-2ab+bb$$

$$(a+b)(a-b) = aa-bb$$

$$(mm-nn)(mm-nn)+4mmnn = (mm+nn)(mm+nn)$$

$$(a+b)(a+b)+(a-b)(a-b) = 2aa+2bb$$

$$(a+b)(a+b)-(a-b)(a-b) = 4ab$$

$$(a+b+c)(a+b-c)(b+c-a)(c+a-b)$$

$$= 2aabb+2bbcc+2ccaa-aaaa-bbbb-cccc$$

$$(a+b)(a+b)(a+b) = aaa+3aab+3abb+bbb$$

$$\frac{1}{a}+\frac{1}{a-1}+\frac{1}{a-2} = \frac{3aa+2-6a}{aaa+2a-3aa}$$

$$\frac{a+b}{a-b}+\frac{a-b}{a+b} = \frac{2aa+2bb}{aa-bb}$$

$$axx+bx+c = \frac{(2ax+b)(2ax+b)+4ac-bb}{4a}$$

$$\frac{xxxx-yyyy}{xxx-yyy} = \frac{xxx+xxy+xyy+yyy}{xx+xy+yy}$$

$$\frac{x+a}{x+b} = \frac{xx+(a+c)x+ac}{xx+(b+c)x+bc}$$

The reduction of several terms into one can be performed *so as to produce a more simple term*, when all the terms are alike as to letters. Thus,

$$3a + 2a = 5a \qquad a + 7a - 4a = 4a$$

$$3ab + 2ab = 5ab \qquad \frac{p}{q} + 7\frac{p}{q} - 4\frac{p}{q} = 4\frac{p}{q}$$

$$3xx + 2xx = 5xx \qquad aab + 7aab - 3aab = 5aab$$

$$a + 12a - 3a - 6a + 2a - a = 5a$$

In the last instance, all the additive terms make up $15a$, from which a is to be subtracted three times, six times, and once, or 10 times in all; that is, $10a$ is to be subtracted: and $15a - 10a = 5a$.

Similarly $a + b - 3a + 4b = 5b - 2a$. For b and $5b$ added give $6b$, which is added to a, after which $3a$ is subtracted. But, subtracting $3a$ where a had previously been added, is the same as subtracting $2a$ without previously adding a, which gives

$$a + 5b - 3a = 5b - 2a$$

EXAMPLES.

$$a + ab - 2ab + 4a + 6a = 11a - ab$$

$$2xx + 6x - 4x - xx + c = xx + 2x + c$$

$$3x - 15 + \tfrac{1}{2}x - x - 7 = 2\tfrac{1}{2}x - 22$$

$$x + y + x - y + 3x = 5x$$

If, upon looking through such an expression as either of the above, we find the following terms containing the simple product xy (with their signs)

$$+6xy, \quad -xy, \quad +4xy, \quad +2xy, \quad -11xy, \quad -12xy,$$

the single term which represents the result of all those containing xy, is found as follows; the whole number of additions of xy to the rest of the expression amounts to 12, and there are 24 subtractions of it in all. Let all the additions, and as many of the subtractions, be abandoned, there will then remain 12 subtractions uncompensated by any additions, and $-12xy$ must be appended to the rest of the expression.

When two terms are not exactly the same as to letters and number of letters which enter them, no such simplification can take place. For example, $aa+a$ cannot be reduced. It is not* $2a$, nor $2aa$, nor aaa; it is a taken a times added to a taken once, and is therefore a taken $a+1$ times, or $(a+1)a$. This is also $(a+1)$ taken a times, or $a(a+1)$, page xi.

Thus $a+a$ is algebraically reducible to the single term $2a$, but $aa+a$ is not so reducible, nor does it admit of any simplification of form, except that which is contained in the formation of its arithmetical value, which cannot be found till we know what number a stands for.

How many times is a contained in $a+b$? This question admits of no answer till we know how many times b contains a; therefore we can only use the algebraical representation $\frac{a+b}{a}$, which has been chosen to signify the number of times which $a+b$ contains a. But let the student observe that this is not an answer to the question, but a method chosen to represent the answer.

How many times is a contained in $ma-na$? Here, though we cannot completely answer this, till we know what m and n stand for, yet the algebraical meaning of ma and na puts our algebraical answer one step nearer to the arithmetical answer, that is, enables us to answer otherwise than by directly writing down $\frac{ma-na}{a}$. For it is evident that when n times is taken away from m times, the remainder is $m-n$ times; that is, $ma-na$ contains a, $m-n$ times.

This must be taken notice of in all algebraical operations. The question, "What is $8a+5a$?" cannot be answered until we know what a stands for; but to "What is the most simple algebraical form of $8a+5a$?" we answer $13a$. And by the algebraical operations of addition, subtraction, multiplication, &c. we mean the methods of changing algebraical expressions into others which are of more simple character. For instance, "to add $a+b$ to $a-b$." The following

$$(a+b)+(a-b)$$

is the first form of the result, derived from representing the thing to be done under algebraical symbols. But its most simple form is $2a$; and in the reduction of the preceding expression to $2a$, consists what

* These are all mistakes to which the beginner is liable.

we shall call algebraical addition. We shall now go through several of these processes.

ADDITION. We wish to add $a+b$ to $c+e$. If to $a+b$ we add c, giving $a+b+c$, we have not added enough, because the quantity to be added is not c, but e more than c. Therefore $a+b+c+e$ is the result, or*

$$(a+b)+(c+e) = a+b+c+e \cdots\cdot (1)$$

To add $c-e$ to $a+b$, we first add c, giving $a+b+c$. But this is adding too much, for e should have been taken from c and the remainder only added. Correct this by taking e from the result, which gives $a+b+c-e$.

$$(a+b)+(c-e) = a+b+c-e \cdots\cdot (2)$$

In the results of (1) and (2) further reduction is impracticable. We shall now try the following.

To add $a+b$ to $a-b$,

$$(a+b)+(a-b) = a+b+a-b = 2a$$

To add $3x-a$ to $2a-x$,

$$(3x-a)+(2a-x) = 3x-a+2a-x = 2x+a$$

To add $ab-b$ to $2ab+c-6b$.

Ans. $2ab+c-6b+ab-b$ or $3ab+c-7b$

From these, and similar operations, we have the following rule of addition:

Write + before the first terms of all the expressions but one, and consider the aggregate of all as one expression. Make the reductions which similarity of terms (as to letters) will allow.

EXAMPLES. $a-b+3c-ab$
$4ab-b+2a-x$
$4x+6a+ab-7$ are to be added.

Answer. $9a-2b+4ab+3c+3x-7$

* When we wish to refer to an equation afterwards, we place a letter or number opposite to it, as is here done.

Add.	Add.	Add.
$a-b$	$a-2b$	$a+ma-4$
$b-c$	$b-2c$	$6-\ \ a+2ma$
$c-d$	$c-2d$	$12ma-12-3a$
$d-x$	$d-2x$	$p+\ a-\frac{1}{2}$
$a-x$	$a-b-c-d-2x$	$15ma-2a-10\frac{1}{2}+p$

The following rule is also derived from (1) and (2):

An expression in brackets preceded by the sign + will not be altered in value if the brackets be struck out.

$$a+(b+c-e) = a+b+c-e$$

SUBTRACTION.—To subtract $b+c$ from a. If we first subtract b, which gives $a-b$, we do not subtract enough, since b should have been increased by c, and the whole then subtracted. Hence c must also be subtracted, giving $a-b-c$, or

$$a-(b+c) = a-b-c \cdots\cdots (3)$$

To subtract $b-c$ from a. If we now subtract b, giving $a-b$, we have subtracted c too much; or $a-b$ is less than the result should be by c. Consequently, $a-b+c$ is the true result, or

$$a-(b-c) = a-b+c \cdots\cdots (4)$$

The following are instances:

$$a-(c-a) = a-c+a = 2a-c$$
$$a-(a-c) = a-a+c = c$$
$$3a+b-(2a-b) = 3a+b-2a+b = a+2b$$
$$a+b-(a-b) = a+b-a+b = 2b$$
$$mx-(q-3mx) = mx-q+3mx = 4mx-q$$

WHEN AN EXPRESSION IN BRACKETS IS PRECEDED BY THE SIGN —, THE BRACKETS MAY BE STRUCK OUT, IF THE SIGNS OF ALL THE TERMS WITHIN THE BRACKETS BE CHANGED, NAMELY, + INTO — AND — INTO +. This is evident from (3) and (4).

As the neglect of this rule is the cause of frequent mistakes, not only to beginners, but to more advanced students, we have printed it thus to attract attention. Still further to impress it on the memory of our younger readers, we beg to inform them that to neglect this rule is the same as declaring that all debts are gains, and all property a loss; that to forgive a debt is to do an injury, and that the more a man is robbed the richer he grows; with a thousand other things of the same kind.

Another proof of the preceding rule is as follows. If we wish to find

$$a-(b+c-p-q) \cdots\cdots\cdots\cdots \text{(A)}$$

we must remember that if two quantities be equally increased, their difference remains the same: thus the difference of $a+x$ and $b+x$ is the same as that of a and b. We are told to diminish a by

$$b+c-p-q$$

which is the same as if we diminished $a+(p+q)$ by

$$(b+c-p-q)+(p+q)$$

or by $\quad b+c-p-q \;+\; p+q \quad$ or by $\quad b+c$

that is, $a-(b+c-p-q)$ is the same as

$$a+p+q-(b+c) \quad \text{or} \quad a+p+q-b-c$$
$$\text{or} \quad a-b-c+p+q \cdots\cdots\cdots\cdots \text{(B)}$$

on comparing (A) and (B) the rule is obvious.

The rule for subtraction is as follows:

Consider the first term of the expression to be subtracted as having the sign +; then change every sign, annex the expression thus changed to the expression which is to be diminished, and make all practicable reductions, as in addition.

From	$a+b-c-x+2z+3ab-14$
Take	$c-2a+x+z-4ab+2\frac{1}{2}$
Rem^r.	$3a+b-2c-2x+z-7ab-16\frac{1}{2}$

From	$a+c$	$x+y-3-a$	$a-b+c-d+e$
Take	$2c-a$	$x-y+3-a$	$a-2b+c+d-e$
Rem^r.	$2a-c$	$2y-6$	$b-2d+2e$

From	$a+b+2c+3d+4e-5f-6g$
Take	$12d+4e-3c+2a+b-g+f$
Rem^r.	$5c-9d-6f-5g-a$

What is

$$a-(b-(c+x))+(b-(x-2b))$$

This, by the rules for expunging brackets, is

$$a-b+(c+x)+b-(x-2b)$$

which, by the same rule, is

$$a-b+c+x+b-x+2b \quad \text{or} \quad a+c+2b$$

Shews that

$$a-\{a-(a-(a-x))\} = x$$

$$a-\{b-(a-(b-x))\} = 2a-2b+x$$

Since all the rules for addition and subtraction are independent of the order in which the terms of the expressions are written, we shall in future not inquire whether an expression is written in a possible or impossible form (page xi), but consider the impossible forms as meaning the same thing as the possible ones. Thus, in $3-7+8$, we shall not regard the subtraction as necessarily to be performed first, and therefore treat the expression as impossible, but we shall consider the order of the operations as immaterial, and the above as $3+8-7$; and the same for other expressions. We could not have done this if the rules for addition and subtraction had required any preliminary inquiry into the possibility of the order of the terms. If we have to subtract the preceding, say from 12, we find that the employment of the impossible form, namely

$$12-(3-7+8) = 12-3+7-8 = 8$$

gives the same result as if the possible form had been employed, as in

$$12-(3+8-7) = 12-3-8+7 = 8$$

Multiplication and Division.—In the multiplication and division of algebraical quantities, it must always be borne in mind that the letters may represent either whole numbers or fractions. We shall first express the rules which have been found in arithmetic for the addition, &c. of fractions which have whole numerators and denominators.

First observe that $\frac{a}{b}$ (when a and b are whole numbers) is the answer to all of the following questions, which are in effect the same.

1. If a unit be divided into b parts, and a of those parts be taken, how many units or parts of units result?

2. How many units or parts of units are there in the bth part of a?

3. How many times, or parts of times, does a contain b? Thus, $\frac{3}{7}$ representing the seventh part of unity repeated three times, is the answer to the questions, "What is the seventh part of three?" and "How many parts of a time does three contain seven?"

It is only the method of speaking, or the idiom of our language, which prevents our explaining in a similar manner the meaning of fractions which have *fractional* numerators and denominators. For instance, if we attempt to explain $\frac{a}{b}$ where a is $2\frac{1}{2}$ and b is $\frac{4}{9}$ in the same way as where a is 3 and b is 7, we shall produce a (yet) unintelligible idiom. Let us suppose some concrete unit, say a mile of length.

What is $\frac{3}{7}$ miles, or $\frac{3}{7}$ of a mile?	What is $\frac{2\frac{1}{2}}{\frac{4}{9}}$ miles?
Cut a mile into 7 equal parts and take 3 of them.	*Cut a mile into $\frac{4}{9}$ equal parts,* and take $2\frac{1}{2}$ of them.

The words in italics are unintelligible, and not having a meaning, may have one given to them.* By altering the manner in which the first of the two preceding is expressed, without changing the meaning, we may find a mode of speech which shall not become unintelligible when $\frac{4}{9}$ is written for 7, and $2\frac{1}{2}$ for 3. As follows:

* The same letter may either stand for a whole number or for a fraction; and we might find out how to group those idioms which belong to whole numbers with those belonging to fractions, so as to be able to pass from one of the first set to the corresponding one of the second. But it will be more convenient to take the modes of speaking which belong to whole numbers, and agree that when fractions are spoken of they shall be used for the corresponding expressions. We have already done this in arithmetic in the word *multiplication,* which originally means, "taking a thing many times," but which we have also made to signify "taking a thing a part of a time." Thus we speak of multiplying by *one half.*

Find the length which taken 7 times gives a mile, and take that length 3 times.

Find the length which taken $\frac{4}{9}$ of a time gives a mile, and take that length $2\frac{1}{2}$ times.

The reduced value of $2\frac{1}{2} \div \frac{4}{9}$ is $\frac{45}{8}$ of a mile, or $5\frac{5}{8}$ miles, and the student may easily shew by the rules of arithmetic that this coincides with the length, $\frac{4}{9}$ of which is a mile, repeated $2\frac{1}{2}$ times. And $5\frac{5}{8}$ is also the answer to the question, "How many times and parts of times does $2\frac{1}{2}$ contain $\frac{4}{9}$?"

It is usual in algebra to make use of modes of speaking with regard to letters, drawn from our notions of whole numbers; but as the letters themselves may stand either for fractions or whole numbers, these modes of speaking are too confined unless they are understood to imply the corresponding modes of speaking, with regard to fractions. For instance, suppose it asked, "What is the value of one acre if x acres cost y pounds?" The answer would always be given as follows:—Divide the number y into x equal parts; then each of those parts is the number of pounds which an acre costs. Thus, if 18 acres cost £36, since 2 is the 18th part of 36, £2 is the price of an acre. But if $\frac{1}{2}$ of an acre cost £$2\frac{1}{3}$, and if we speak of dividing $2\frac{1}{3}$ into $\frac{1}{2}$ equal parts, we must mean the same as if we said "take $2\frac{1}{3}$ *two* times," or divide $2\frac{1}{3}$ by $\frac{1}{2}$ (according to the arithmetical rule). Now, though it may appear at first sight almost ridiculous to say that dividing a quantity into $\frac{1}{10}$ equal parts is the same thing as taking it 10 times, we must remember, 1st. that we have done something of the same sort, when we said that dividing into 10 equal parts, and taking one part, is *multiplying* by $\frac{1}{10}$; 2d. that we do say this, when we say that if x acres cost y pounds, the price of one acre is found by dividing y into x equal parts; for letters may stand either for fractions or whole numbers.

We should recommend the student to solve various simple

questions in the rule of three, with fractional *data*.* 1st. By imitation of a question with integral *data*. 2d. By independent reasoning; as follows:

I. If 6 yards cost 7 shillings, how much do 5 yards cost?

As 6 yards to 5 yards, so are 7 shillings to $\frac{7\times 5}{6}$ or $5\frac{5}{6}$ shillings, the answer.

If $\frac{2}{3}$ of a yard cost $\frac{5}{7}$ of a shilling, how much do $\frac{4}{9}$ of a yard cost?

As $\frac{2}{3}$ of a yard to $\frac{4}{9}$ of a yard so is $\frac{5}{7}$ of a shilling to $\frac{\frac{5}{7}\times\frac{4}{9}}{\frac{2}{3}}$ or $\frac{10}{21}$ of a shilling, the answer.

II. If $\frac{2}{3}$ of a yard costs $\frac{5}{7}$ of a shilling

2 yards cost $\frac{5}{7}\times 3$ or $\frac{15}{7}$ ......

1 yard costs $\frac{15}{7}\div 2$ or $\frac{15}{14}$ ......

4 yards cost $\frac{15}{14}\times 4$ or $\frac{30}{7}$ ......

$\frac{4}{9}$ of a yard costs $\frac{30}{7}\div 9$ or $\frac{10}{21}$ ......

We shall now write all the rules of arithmetic relative to fractions in an algebraical form, these rules being those with which the student is already familiar as applied to whole numbers. We shall first suppose the letters to denote whole numbers.

$$\frac{a}{b}=\frac{ma}{mb} \qquad \frac{a}{b}+\frac{c}{d}=\frac{ad+bc}{bd}$$

$$a+\frac{b}{c}=\frac{ac+b}{c} \qquad \frac{a}{b}-\frac{c}{d}=\frac{ad-bc}{bd}$$

$$a-\frac{b}{c}=\frac{ac-b}{c} \qquad \frac{a}{b}-c=\frac{a-bc}{b}$$

* *Datum*. A thing given, that is, a number given as part of a question. "2 pints cost 4 shillings, how much, &c.;" here the *data* are 2 pints and 4 shillings.

$$\frac{a}{b} \times c = \frac{ac}{b} \qquad \frac{a}{b} \times \frac{c}{d} = \frac{ac}{bd}$$

$$\frac{a}{b} \div c = \frac{a}{bc} \qquad c \div \frac{a}{b} = \frac{bc}{a}$$

$$\frac{a}{b} \div \frac{c}{d} = \frac{ad}{bc} = \frac{a}{b} \times \frac{d}{c}$$

The learner must make himself acquainted with each of these equations, which he will easily do if he have the requisite degree of familiarity with the operations of fractional arithmetic. For example, in $\frac{a}{b} \times \frac{c}{d} = \frac{ac}{bd}$, he will recognise the rule, "To multiply one fraction by another, multiply their numerators for a numerator, and their denominators for a denominator."

The preceding rules all hold good when the letters stand for fractions. We shall take the demonstration of this in one instance, namely, $\frac{a}{b} = \frac{ma}{mb}$. Let a be the fraction $\frac{p}{q}$ let b be $\frac{r}{s}$, and let m be $\frac{x}{y}$; where p, q, r, s, x, and y, are whole numbers.

$$\frac{a}{b} = \frac{\frac{p}{q}}{\frac{r}{s}} = \frac{ps}{qr} \quad \text{(By the rule.)}$$

$$ma = \frac{x}{y} \times \frac{p}{q} = \frac{xp}{yq} \qquad mb = \frac{x}{y} \times \frac{r}{s} = \frac{xr}{ys}$$

$$\frac{ma}{mb} = \frac{\frac{xp}{yq}}{\frac{xr}{ys}} = \frac{xyps}{xyqr}$$

$$\text{But } \frac{xyps}{xyqr} = \frac{(xy) \times ps}{(xy) \times qr} = \frac{ps}{qr}$$

$$\text{or} \quad \frac{ma}{mb} = \frac{a}{b} \quad \left.\right\} \text{where } a,\ b, \text{ and } m, \text{ are fractional.}$$

It is shewn in arithmetic that the order in which multiplications or divisions are made may be changed without affecting the result, both in questions of whole numbers and fractions. We need not, when speaking of both, distinguish between multiplication and division; for division by a is multiplication by $\frac{1}{a}$. The following summary should be attended to and repeated on several products.

The expression $abcd$ is the product, 1. Of a, b, c, and d, in any order; 2. Of ab and cd; 3. Of ac and bd; 4. Of ad and bc; 5. Of abc and d; 6. Of abd and c; 7. Of acd and b; 8. Of bcd and a.

The product of single terms may be expressed by writing the letters consecutively in any order which may be convenient, and actually multiplying the numerical coefficients, if any. The sign $\times$ becomes necessary between two numerals only. Thus, $2ab \times 4cd$ might be written in the following ways:

$$2ab4cd \qquad 2\times 4abcd \qquad 8acbd \qquad 8abcd, \text{ \&c.}$$

It is most convenient to write the numerical coefficient first, and letters of the same sort together, the order of the alphabet being generally preferred. Thus,

$$2aab \times 3abbc \quad \text{is written} \quad 6aaabbbc$$

$$12abx \times 4abxx = 48aabbxxx \quad 3abc \times \frac{1}{2}ab = \frac{3}{2}aabbc$$

When fractions are written side by side of each other, or of single letters, multiplication is intended to be denoted. Thus,

$$2\frac{a}{b}c\frac{e}{f} \quad \text{means} \quad 2\times\frac{a}{b}\times c\times\frac{e}{f} \quad \text{and is} \quad \frac{2ace}{bf}$$

[Though a, ab, abc, &c. are called *integral* expressions, and $\frac{a}{b}$, $\frac{c}{d}$, $\frac{ac}{b}$, &c. *fractions*, this refers to their algebraical, not to their arithmetical character; for, since letters may stand for fractions, that which is integral considered algebraically, may be fractional considered arithmetically, and *vice versâ*. For example, suppose that a stands for $\frac{1}{2}$ and b for $\frac{1}{4}$, then the algebraically integral expression a is the arithmetical fraction $\frac{1}{2}$; while the algebraical fraction $\frac{a}{b}$, or $\frac{1}{2} \div \frac{1}{4}$ is the arithmetical whole number 2.

In speaking, therefore, of integral or fractional expressions, we refer always to their algebraical appearance, not to their arithmetical values. The latter depend on the particular arithmetical value of the letters, on which no supposition is made.*]

* The part of algebra which treats of letters considered as representing *whole numbers* only, is generally called the *theory of numbers*. It is very rarely, if ever, of use in the application of algebra to the Differential Calculus, and is therefore altogether omitted in this work.

The multiplication of algebraical quantities depends on the rule which appears in the following equations:

$$m(a+b) = ma+mb$$
$$m(a-b) = ma-mb$$

First, it is required to take $a+b$, m times. If a be taken m times, it is clear that for every time and part of a time which a has been taken, b or a part of b has been omitted. Consequently, ma is too little by mb, or $ma+mb$ is the product required.

Secondly, it is required to take $a-b$, m times. Here, if a be taken m times, b or a part of b too much has been taken for every time or part of a time which a has been taken. Consequently, ma is too much by mb, or $ma-mb$ is the product required.

Hence we may prove the following equation:

$$m(a+b-c-d) = ma+mb-mc-md$$

as follows: let $a+b$ be called p, and let $c+d$ be called q; then $a+b-c-d$ is $p-q$ (page xix.), and

$$m(a+b-c-d) = m(p-q) = mp-mq$$

But $\quad mp = ma+mb \qquad mq = mc+md$

$$mp-mq = ma+mb-(mc+md)$$
$$= ma+mb-mc-md$$

The following are applications of the preceding rules:

$$3(a+b) = 3a+3b \qquad 3\frac{1}{2}(a-b) = 3\frac{1}{2}a-3\frac{1}{2}b$$

$$ab(a-b) = aab-abb \qquad 2a(a-aa)^* = 2aa-2aaa$$

$$3abc(ab-ac+4) = 3aabbc-3aabcc+12abc$$

$$2\left(\frac{x}{2}+\frac{x}{3}\right) = x+\frac{2x}{3} \qquad 6\left(\frac{x}{2}+\frac{x}{3}\right) = 3x+2x$$

* The student will observe, that in these examples we do not inquire whether the expressions are possible or not, but how to apply the rules when they *are* possible. For instance, $a-aa$ is impossible when a is greater than 1, and possible only when a is less than 1. And let it be observed in proceeding, as a matter of great importance in the subsequent part of the work, that *the rules investigated will not serve to distinguish between possible and impossible expressions.*

$$40\left(\frac{1}{2}-x\right) = 20-40x \qquad a\left(\frac{3}{4}+b\right) = \frac{3}{4}a+ab$$

$$\frac{a}{b}\left(\frac{b}{a}+b\right) = \frac{a}{b}\,\frac{b}{a}+\frac{a}{b}b = 1+a$$

$$\frac{xy}{z}\left(xz+1\right) = \frac{xxyz}{z}+\frac{xy}{z} = xxy+\frac{xy}{z}$$

$$\frac{a}{a+b}\left\{c+d-e\right\} = \frac{ac}{a+b}+\frac{ad}{a+b}-\frac{ae}{a+b}$$

$$pqrs\left(\frac{1}{p}+\frac{1}{q}+\frac{1}{r}-\frac{1}{pqr}\right) = qrs+prs+pqs-s$$

In order to multiply $a+b$ by $c+d$, let us, for a moment, make p stand for $a+b$. Then, p multiplied by $c+d$, is the same as $c+d$ multiplied by p, which is $pc+pd$, or

$$(a+b)\,(c+d) = (a+b)c+(a+b)d$$

But $$(a+b)c = ac+bc, \qquad (a+b)d = ad+bd$$

$$\therefore\ (a+b)\,(c+d) = (ac+bc)+(ad+bd)$$
$$= ac+bc+ad+bd$$

To multiply $a+b$ by $c-d$, let $a+b=p$, and we have $p(c-d)=pc-pd$, or

$$(a+b)\,(c-d) = (a+b)c-(a+b)d$$
$$= (ac+bc)-(ad+bd)$$
$$= ac+bc-ad-bd$$

To multiply $a-b$ by $c-d$, let $a-b=p$, and we have $p(c-d)=pc-pd$, or

$$(a-b)\,(c-d) = (a-b)c-(a-b)d$$
$$= (ac-bc)-(ad-bd)$$
$$= ac-bc-ad+bd$$

There are two methods of conducting this process, the first of which is recommended to the learner for the present.

1. To multiply $a+b-2c$ by $d-a-c$.

Here a times and c times the multiplicand are to be successively taken from d times the multiplicand; that is, $a+c$ times is to be taken from d times.

d times the multiplicand $= ad+bd-2cd$

Add $\begin{cases} a \text{ times ditto} = aa+ab-2ac \\ c \text{ times ditto} = ac+bc-2cc \end{cases}$

$a+c$ times ditto $= aa+ac+ab+bc-2ac-2cc$

$\begin{cases} \text{Subtract from } d \text{ times} & ad+bd+2ac+2cc-2cd \\ \text{ditto which gives} & -aa-ac-ab-bc \end{cases}$

2. From looking at the preceding instances, it appears that the rule of multiplication is as follows: *Consider the first terms as having the sign* + ; multiply *every term of the multiplicand by every term of the multiplier, and put* + *before the products of terms which have the same sign, and* — *before the products of terms which have different signs.* The preceding example is here written in the usual way, with the proper signs written to every term by the preceding rule.

$$\begin{array}{lll} & a+b-2c & \\ & d-a-c & \\ \hline \text{from } d \cdots & ad+bd-2cd & \\ \text{from } a \cdots & -aa-ab+2ac & \text{Product required.} \\ \text{from } c \cdots & -ac-bc+2cc & \end{array}$$

Where like terms are found in different lines, so that subsequent reductions may be made, it is convenient to put like terms under each other, as in the following example:

$$\begin{array}{lll} \text{Multiply} & xx-2x+1 & \\ \text{By} & x-4 & \\ \hline \text{From } x & xxx-2xx+x & \text{Product required.} \\ \text{From } 4 & -4xx+8x-4 & \\ \hline & xxx-6xx+9x-4 & \text{ditto in simplest form.} \end{array}$$

But the manner of doing this, which (as yet) we recommend, is the following:

$$\begin{array}{lll} \text{Multiply} & xx-2x+1 & \\ \text{By} & x-4 & \\ \hline \text{From } x \text{ times multiplic}^{d} & xxx-2xx+x & \text{Subtract second} \\ \text{Take } 4 \text{ times ditto} & 4xx-8x+4 & \text{line from first, as in third line.} \\ \hline & xxx-6xx+9x-4 & \text{Product required.} \end{array}$$

The three instances which follow are of particular importance, and the learner should be able to write them (and similar ones) at sight.

Mult.	$a+b$		$a-b$		$a+b$
By	$a+b$		$a-b$		$a-b$
To	$aa+ab$	From	$aa-ab$	From	$aa+ab$
Add	$ab+bb$	Take	$ab-bb$	Take	$ab+bb$
	$aa+2ab+bb$		$aa-2ab+bb$		$aa \qquad -bb$

[The following definitions may be appropriately introduced here.

A *square* is a four-sided figure, with sides of equal length, and with contiguous sides perpendicular to each other.

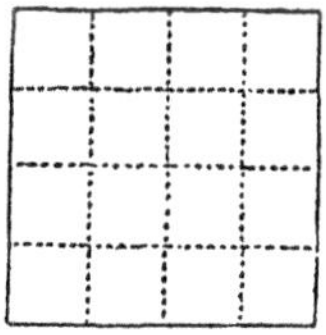

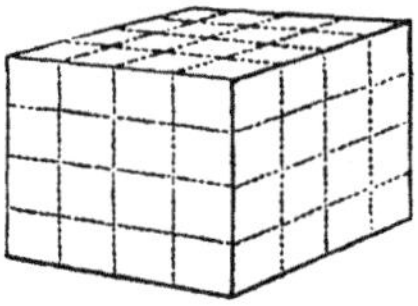

A cube is a solid figure enclosed by six equal squares; or, a box of the same length, breadth, and thickness.

A square 4 inches long contains 4×4 squares of one inch long; a square of x inches long contains xx squares of one inch long. There are x rows of square inches, and x squares in each row.

A cube 4 inches long contains $4 \times 4 \times 4$ cubes of one inch long; a cube x inches long contains xxx cubes of one inch long. There are x layers of cubic inches, and xx cubic inches in each layer.

Owing to this connexion of xx with the square on a line of x units, and of xxx with the cube on a line of x units, it has always been customary to call xx the *square* of x, and xxx the *cube* of x. But $xxxx$ is called the *fourth power* of x, $xxxxx$ the *fifth power*; and so on. So that x itself should be called the *first power* of x, xx the *second power* of x, and xxx the *third power* of x. But the words square and cube are so conveniently short that they have never been abandoned.]

The last mentioned products may be thus stated:

1. $$(a+b)(a+b) = aa + 2ab + bb$$

or, *The square of the* sum *of two quantities is the* sum *of their squares,* augmented *by twice their product.*

2. $$(a-b)(a-b) = aa - 2ab + bb$$

or, *The square of the* difference *of two quantities is the* sum *of their squares,* diminished *by twice their product.*

3. $$(a+b)(a-b) = aa - bb$$

or, *The* sum *of two quantities, multiplied by their* difference, *is the* difference *of their squares.*

Thus let the two quantities be ab and $2a$.

$$ab \times ab = aabb \qquad 2a \times 2a = 4aa$$

$$2(ab \times 2a) = 4aab$$

$$(ab+2a)(ab+2a) = aabb+4aab+4aa$$

$$(ab-2a)(ab-2a) = aabb-4aab+4aa$$

$$(ab+2a)(ab-2a) = aabb-4aa$$

The following are examples for the student:

Square* of $\left(a \pm \frac{1}{a}\right) = aa \pm 2 + \frac{1}{aa}$

$$\left(a+\frac{1}{a}\right)\left(a-\frac{1}{a}\right) = aa - \frac{1}{aa}$$

Square of $(2ax \pm b) = 4aaxx \pm 4abx + bb$

$$(2ax+b)(2ax-b) = 4aaxx - bb$$

Square of $(a+b+c) = (a+b)(a+b) + 2(a+b)c + cc$

$$= aa + bb + cc + 2ab + 2bc + 2ca$$

$$(a+b+c)(a+b-c) = (a+b)(a+b) - cc$$

$$= aa + bb - cc + 2ab$$

$$(c+a-b)(b+c-a) = (c+\overline{a-b})(c-\overline{a-b})$$

* By $\pm$ occurring several times in an equation, two equations are combined in one. The upper or under sign is to be taken throughout. Thus,

$a \pm b = c \mp d$ is either

$a+b=c-d$ or $a-b=c+d$

The student is recommended never to use this double sign himself; but as it frequently occurs in books it is here shewn.

$$= cc-(a-b)(a-b)$$
$$= 2ab+cc-aa-bb$$
$$= 2ab-(aa+bb-cc)$$

From the last two examples, shew that the product of the four quantities

$$a+b+c, \quad a+b-c, \quad b+c-a, \quad c+a-b, \text{ is}$$
$$2aabb+2bbcc+2ccaa-aaaa-bbbb-cccc$$

Shew the preceding by help of the following (which shew also),

Square of $(p+q-r) = pp+qq+rr+2pq-2qr-2rp$

The following are miscellaneous examples in multiplication, *to be done without paper:*

$$(a+bx)(a+cx) = aa+abx+acx+bcxx$$
$$(x+a)(x+b) = xx+ax+bx+ab$$
$$(x-a)(x-b) = xx-ax-bx+ab$$
$$(x+1)(x-3) = xx-2x-3$$
$$(x-1)(x-3) = xx-4x+3$$
$$(2x+1)(x-1) = 2xx-x-1$$

The following theorems are given for exercise :

1. If a and b be two quantities, of which a is the greater, and if S be the square of their sum, D the square of their difference, and P the product of their sum and difference, then

$$S+D = 2(aa+bb) \qquad S-D = 4ab$$
$$S+P = 2a(a+b) \qquad S-P = 2b(a+b)$$
$$D+P = 2a(a-b) \qquad P-D = 2b(a-b)$$

2. If two numbers differ by a unit, the difference of their squares is their sum; and if two fractions are together equal to a unit $\left(\text{such as } \frac{1}{4} \text{ and } \frac{3}{4}\right)$ their difference is also that of their squares.

3. The sum of the squares of $xx-yy$ and $2xy$ is the square of $xx+yy$.

[According to the rule, the square of $a-b$ is the same as the square of $b-a$; for that of the first is $aa-2ab+bb$, and that of the second $bb-2ab+aa$, which two are the same (see page xi).

But one of the two, $a-b$ or $b-a$, must be impossible, except only when $b=a$, in which case both are nothing; for in every other case either $a-b$ or $b-a$ is an attempt to subtract the *greater* from the *less*. But for the same reason, one of the two, $b-a$ or $a-b$, must be possible; therefore $aa+bb-2ab$, which is the square of that possible one, must also be possible. That is, if either of the two, a or b, exceed the other, $aa+bb$ must be greater than $2ab$. And we also see that *it does not follow that an algebraical process is intelligible because it gives an intelligible result;* for it appears that the algebraical rule of multiplication would apply to (3 — 7) (3 — 7), which is absurd, and give the same result as (7 — 3) (7 — 3), or 16. This is a defect, the remedy for which we shall afterwards have to find; and we see, that so far as we have yet gone, we can never know that any process is correct which is to lead to the value of an unknown quantity, until we re-examine the process after the unknown quantity has been found by it.]

Divisions in algebra we shall for the present divide into two classes; those which it is obvious how to do, and those which it is not obvious how to do. For instance, to divide ab by a, that is, to tell how many times ab contains a, the answer evidently is, that since ab means the same as ba, or *a taken b times*, therefore ab must contain a, b times; or ab divided by a is b. In this case the simplest rule is as follows: *To divide, where there has already been a multiplication by the quantity which is made a divisor, suppress all the symbols of the multiplication.*

This will be seen in the following examples:

Dividend.	Divisor.	Quotient.	Dividend.	Divisor.	Quotient.
ab	a	b	$12aax$	$6aax$	2
abc	ab	c	$2\frac{1}{2}byz$	$\frac{1}{4}by$	$10z$
$2abx$	$2x$	ab	$aaaa$	aaa	a
$aabbx$	ab	abx	$aaaa$	aa	aa
$6abcc$	$3abc$	$2c$	xyz	xyz	1

[One of the most common errors of a beginner is a mistake between 0 and 1, arising from a confusion of subtraction and division. This is partly a result of the idiom of our language, as follows. If a beginner be asked how many *times* does 7 contain 7, the answer is sure to be, *no times at all;* and in one sense this is correct, for 7 does not contain 7 a number of *times*, but *one time.* But it must always be understood in algebra that times means time, or times, or parts of a time, or time and parts of a time, or times and parts of a time.* Therefore, though

x *diminished by* x *leaves nothing,*
x *divided by* x *gives one.*]

Closely connected with the preceding is the theorem in fractions that $\frac{a}{b}$ and $\frac{ma}{mb}$ are the same. For $\frac{a}{b}$ multiplied by m gives $\frac{ma}{b}$, and this divided by m gives $\frac{ma}{mb}$. But multiplication, followed by division (multiplier and divisor being the same), leaves any quantity the same as at first.

From the rule of multiplication it follows, that if any quantity contain the same letter or letters in every term, it is the obvious result of multiplying another expression by that letter or the product of those letters. Thus, $ab+ac$ is $b+c$ multiplied by a, $aab-abc$ is $a-c$ multiplied by ab. Hence, to divide an expression by a letter or a product of letters, strike out those letters from every term of the dividend. But remember to write 1 where all the letters of any term are thus struck out. For instance, $a+ab$ divided by a gives $1+b$, $ac-aac+acc$ divided by ac gives $1-a+c$.

The following are instances, arranged as before:

Dividend.	Divisor.	Quotient.
$2ab-2bc+4abc$	$2b$	$a-c+2ac$
$aaa-aa+a$	a	$aa-a+1$
$6ab-3a+3b$	3	$2ab-a+b$
$aab-abb$	ab	$a-b$
$axxy-xxxyy$	xxy	$a-xy$

* An act of parliament, or any other legal instrument, always speaks of men under the title "man or men," &c. If this should happen to be neglected, and a single offender should plead that "men" only were prohibited from doing as he (one man) had done, it would be called a "quibble." If the student, after this warning, should ever say that x contains x *no times*, it would not only be a quibble, but a very useless quibble, because nothing is to be got by it.

In dividing ab by a we might proceed as follows. The result of the process is the fraction $\frac{ab}{a}$, which is not changed in value if both numerator and denominator be divided by a. But this gives $\frac{b}{1}$, which is b.

Such a process is of no use in the preceding case; but suppose that ab is to be divided by ac. The complete division is here impossible until we know what numbers a, b, and c, stand for. But $\frac{ab}{ac}$, the symbol of the result, may be reduced to $\frac{b}{c}$ by the preceding theorem. The division here is not completed, but reduced to a more simple division.

$$\frac{2b}{2c} = \frac{b}{c} \qquad \frac{aab}{ax} = \frac{ab}{x} \qquad \frac{3ammn}{6aam} = \frac{mn}{2a}$$

$$\frac{p}{pq} = \frac{1}{q} \qquad \frac{3ab}{aab} = \frac{3}{a} \qquad \frac{21vww}{28xw} = \frac{3vw}{4x}$$

The division of an expression of several terms by another of one term may also admit of reductions. For example, $xy + yz - zx$ divided by xyy is

$$\frac{xy}{xyy} + \frac{yz}{xyy} - \frac{zx}{xyy} \quad \text{or} \quad \frac{1}{y} + \frac{z}{xy} - \frac{z}{yy}$$

$$\frac{2v - xx + vx}{vx} = \frac{2}{x} - \frac{x}{v} + 1$$

$$\frac{a + b + c}{abc} = \frac{1}{bc} + \frac{1}{ac} + \frac{1}{ab}$$

$$\frac{a + b}{aa} = \frac{1}{a} + \frac{b}{aa} \qquad \frac{aa + 1}{a} = a + \frac{1}{a}$$

$$\frac{x + 4y - 3z + 2}{6} = \frac{x}{6} + \frac{2y}{3} - \frac{z}{2} + \frac{1}{3}$$

All that precedes contains the obvious cases of division; of those, the answer to which requires further process, the following is an instance: "How often is $x + y$ contained in $xxx + yyy$?" As we shall not need such a process for some time, we defer it to its proper place. In some cases, however, the preceding theorems of multiplication (page xxix) furnish an answer at once. For instance, we know that $xx - 9$ must contain $x + 3$, $x - 3$ times.

Fractions may frequently be reduced to simpler terms by inspection, of which the following are instances:

$$\frac{a+ab}{a-ab}=\frac{1+b}{1-b} \qquad \frac{3x+6xx}{9xy-3x}=\frac{1+2x}{3y-1}$$

$$\frac{a-aa}{2a+az}=\frac{1-a}{2+z} \qquad \frac{aa+3ab}{ab+12a}=\frac{a+3b}{b+12}$$

It is frequently necessary to arrange expressions in a different form, without altering their value, by performing inverse operations upon them with the same *data*, such as addition followed by subtraction, or multiplication followed by division. The four following methods of writing x exhibit this process.

$$x+a-a \qquad x-a+a \qquad \frac{ax}{a} \qquad \frac{x}{a}\times a$$

Thus $a+x=2a+x-a=\left(1+\frac{x}{a}\right)a$

$$aa+2ab-c=aa+2ab+bb-(c+bb)$$
$$=(a+b)(a+b)-(c+bb)$$

$$bb-4ac=bb\left(1-\frac{4ac}{bb}\right)=abc\left(\frac{b}{ac}-\frac{4}{b}\right)$$

$$m+n=mn\left(\frac{1}{n}+\frac{1}{m}\right)=n\left(\frac{m}{n}+1\right)=m\left(1+\frac{n}{m}\right)$$

$$x=\frac{1}{\frac{1}{x}}=\frac{1+x}{\frac{1}{x}(1+x)}=\frac{1+x}{\frac{1}{x}+1}$$ (See page v.)

The following are instances of those reductions of fractions which will occur hereafter. The rules with regard to fractions which are proved in arithmetic are here applied in conjunction with the algebraical methods of addition, &c. At the head of each section stands an example without any complicated expressions, containing the arithmetical process used in those which succeed.

I. $$\frac{a}{b}=\frac{ma}{mb}$$

$$\frac{1+\frac{1}{x}}{1+\frac{1}{xx}}=\frac{\left(1+\frac{1}{x}\right)xx}{\left(1+\frac{1}{xx}\right)xx}=\frac{xx+x}{xx+1}$$

$$\frac{v+\frac{1}{2}}{\frac{2}{3}v+\frac{1}{4}}=\frac{\left(v+\frac{1}{2}\right)12}{\left(\frac{2}{3}v+\frac{1}{4}\right)12}=\frac{12v+6}{8v+3}$$

$$y = \frac{y(y-1)}{y-1} = \frac{yy-y}{y-1} \qquad \frac{x}{a} = \frac{xx+ax}{ax+aa}$$

$$\frac{1}{a+b} = \frac{a-b}{aa-bb} = \frac{a+b}{aa+2ab+bb} = \frac{4}{4a+4b}$$

$$\frac{x-4}{2\frac{1}{2}} = \frac{2x-8}{5} = \frac{4x-16}{10} = \frac{2ax-8a}{5a}$$

$$\frac{7x-4}{10} = \frac{\frac{1}{2}(7x-4)}{\frac{1}{2}(10)} = \frac{3\frac{1}{2}x-2}{5}$$

$$\frac{\frac{1}{a}+\frac{1}{ab}}{b-a+\frac{1}{b}} = \frac{ab\left(\frac{1}{a}+\frac{1}{ab}\right)}{ab\left(b-a+\frac{1}{b}\right)} = \frac{b+1}{abb-aab+a}$$

II. $a+\frac{x}{y} = \frac{ay+x}{y} \qquad a-\frac{x}{y} = \frac{ay-x}{y} \qquad \frac{x}{y}-a = \frac{x-ay}{y}$

$$1-\frac{1}{x} = \frac{x-1}{x} \qquad x-\frac{1}{x} = \frac{xx-1}{x}$$

$$2-\frac{y-1}{y+1} = \frac{2y+2-(y-1)}{y+1} = \frac{y+3}{y+1}$$

$$a-\frac{aa}{a+b} = \frac{ab}{a+b} \qquad a-\frac{ab}{a+b} = \frac{aa}{a+b}$$

$$a+b-\frac{aa-2ab}{a+b} = \frac{4ab+bb}{a+b}$$

$$a+b+\frac{aa-2ab}{a+b} = \frac{2aa+bb}{a+b}$$

$$\frac{y+1}{y-1}+y+1 = \frac{y+yy}{y-1} = y\frac{y+1}{y-1}$$

$$\frac{a+b-c}{a-c}-2 = \frac{b-a+c}{a-c} \qquad \frac{x}{y}-x = \frac{x-xy}{y}$$

$$\frac{ab+bc+ca}{a+b+c}-c = \frac{ab-cc}{a+b+c}$$

III. $\frac{a}{b}+\frac{x}{y} = \frac{ay+bx}{by} \qquad \frac{a}{b}-\frac{x}{y} = \frac{ay-bx}{by}$

$$\frac{a+b}{a-b}-\frac{a-b}{a+b} = \frac{(a+b)(a+b)-(a-b)(a-b)}{(a-b)(a+b)} = \frac{4ab}{aa-bb}$$

$$\frac{1}{x}+\frac{1}{y} = \frac{x+y}{xy} \qquad \frac{x}{y}+\frac{y}{x} = \frac{xx+yy}{xy}$$

$$\frac{a+b}{c+d}-\frac{a}{c} = \frac{cb-ad}{cc+cd} \qquad \frac{a-b}{c-d}-\frac{a}{c} = \frac{ad-bc}{cc-cd}$$

$$\frac{x}{x+y} - \frac{y}{x-y} = \frac{xx-2xy-yy}{xx-yy} \qquad \frac{p}{q} + \frac{q}{pp} = \frac{ppp+qq}{ppq}$$

$$\text{IV.} \quad \frac{a}{b} \times \frac{x}{y} = \frac{ax}{by} \qquad \frac{a}{b} \div \frac{x}{y} = \frac{ay}{bx}$$

$$\frac{x-1}{x+1} \times \frac{x+2}{x-1} = \frac{(x-1)(x+2)}{(x+1)(x-1)} = \frac{x+2}{x+1}$$

$$\frac{2ab}{a+b} \div \frac{a-b}{3a} = \frac{6aab}{aa-bb} \qquad \frac{3ax}{yy} \times \frac{yv}{2x} = \frac{3av}{2y}$$

$$\frac{m}{an} \div \frac{2m}{3bn} = \frac{3b}{2a} \qquad \frac{pc}{q} \times \frac{3qq}{cce} = \frac{3pq}{ce}$$

The student is recommended to make himself well acquainted with every example given in the preceding list, *but no more* (see the Preface); as a better method of obtaining examples will be given.

All that has preceded is purely arithmetical, and the letters may be considered as mere abbreviations of numbers, and all identical equations as abbreviations of arithmetical propositions. Thus,

$$(a+b)(a+b) = aa + 2ab + bb$$

represents the following sentence:—If two numbers be added together, and if the sum be multiplied by itself, the result is the same as would arise from multiplying each number by itself, and adding to the sum of these products twice the product of the numbers.

An *arithmetical* problem is one in which numbers are given, and certain operations; and the question asked is, what number will result from performing the given operations upon the given numbers. For instance, what is the fiftieth part of the product of 25 and 300.

An *algebraical* problem is one in which numbers are either given or supposed to be given (as will presently be further explained), and a question is asked of which it is not at once perceptible *what operations will furnish the answer.* Such is the following:—The numbers 3 and 17 are given; what number is that, the double of which will fall short of 17 by as much as its half exceeds 3? And the questions asked are the following. 1. Is there any such number? 2. If there be, by what operations on 3 and 17 may it be found? 3. What is the result of these operations, or the number required. The answers to which (as the student may afterwards find) will be that there is such a number, that it is found by taking two-fifths of the sum of 3 and 17, and that in consequence the number is 8.

If we had contented ourselves with the first two questions, it would have been unnecessary to have specified that the numbers in question were 3 and 17, for the same problem might have been proposed about any other numbers, and the process of solution would (as may afterwards be shewn) have been the same whatever the numbers might have been. That is, if the following question had been asked;—The numbers a and b are given; what number is that, the double of which will fall short of b (the greater), as much as its half exceeds a (the less)? The answer is $\frac{2}{5}(a+b)$ and the verification is as follows:

$$\text{The double of } \frac{2}{5}(a+b) \text{ is } \frac{4}{5}(a+b), \text{ or } \frac{4}{5}a+\frac{4}{5}b.$$

$$\text{This falls short of } b \text{ by } b-\left(\frac{4}{5}a+\frac{4}{5}b\right) \text{ or } b-\frac{4}{5}a-\frac{4}{5}b$$

$$\text{or } \frac{1}{5}b-\frac{4}{5}a; \text{ but the half of } \frac{2}{5}(a+b) \text{ is } \frac{1}{5}(a+b)$$

which exceeds a by $\frac{1}{5}(a+b)-a$, or $\frac{1}{5}a+\frac{1}{5}b-a$, or $\frac{1}{5}b-\frac{4}{5}a$, the same as that by which the double of $\frac{2}{5}(a+b)$ falls short of b. Which was to be done. Now, observe that the preceding not only informs us of the general process by which this problem may be solved, but it also shews in what cases the problem is impossible. For the excess or defect above-mentioned turns out to be $\frac{1}{5}b-\frac{4}{5}a$, which is an absurdity, unless $\frac{1}{5}b$ be greater than (or at least not less than) $\frac{4}{5}a$; that is, unless b be greater than $4a$. This was the case in the first instance where b was 17 and a was 3. If it be not so, we may pronounce that the problem is impossible; for instance, let the student try to find a number or fraction, the double of which shall fall short of 11 by as much as its half exceeds 3. In this problem there must be a contradiction; and when we know there is one, and set ourselves to find out how it arises, we see it in the following:—The number sought is presumed to have a half which exceeds 3 (so that it must be more than 6) and a double which falls short of 11 (so that it must be less than $5\frac{1}{2}$). But a number which exceeds 6 cannot be less than $5\frac{1}{2}$; therefore the clauses of the preceding problem contradict each other.

We see, then, that we may propose a problem which is impossible or contradictory, or has no solution. But, on the other hand, we may propose a problem which admits of an innumerable number of answers. These we will call *unlimited* problems. And, as in the case of impossible problems, there are some of which the impossibility is evident, as "To find a whole number which shall be the half of seven;" and others, in which it requires investigation to discover the impossibility, as "To divide 10 into two parts, whole or fractional, of which the product shall be 30;" so in the case of unlimited problems, there are those in which the unlimited nature of the result shall be evident, and others in which it shall not be so. For instance, to the question, "To find two odd numbers which added together shall make an even number?" it is clear that the answer is, "*Any* two odd numbers;" and to the question, "What two numbers are those of which half the sum added to half the difference shall give the greater number?" the answer is (but not so evidently), "*Any* two numbers." Between *these* two extremes, we can conceive there may be problems which admit of 1000 answers, others of 999 answers, &c. &c. down to problems which admit only of one answer. And even when we find that a problem is impossible, we may yet think proper to ask, *why* is it impossible? what are the two parts of the problem which contradict each other, and by how much? that is to say, what sort of change, and quantity of change, in the conditions of the problem, will render it possible?

To all these questions, arithmetic gives no means of answering, and we have therefore to consider algebra as a distinct science, which proposes objects of which arithmetic knows nothing, and therefore as we may suppose, uses language, finds methods, and adopts interpretations, of which arithmetic furnishes no examples.

If the student have read* a little of geometry (a science which he

* In England, the geometry studied is that of Euclid, and I hope it never will be any other; were it only for this reason, that so much has been written on Euclid, and all the difficulties of geometry have so uniformly been considered with reference to the form in which they appear in Euclid, that Euclid is a better key to a great quantity of useful reading than any other.

should begin to study at the same time as algebra, if not before), he knows that all the questions of geometry are made for him, that is, the reasonings, &c. are put together before his eyes, and all he has to do is to comprehend and agree to one step of the process after another. This is called *synthesis* (*συνθεσις, a putting together*), or the *synthetical* method, in opposition to *analysis* (*ἀναλυσις*, an unloosing, or bringing asunder), or the *analytical* method. The latter consists in taking the problem to pieces, if the phrase may be used, that is, reasoning upon the whole problem, reducing it to more and more simple terms, and so coming at last to those considerations which must be put together to make a solution and to verify it.

We now proceed to establish the principles of algebra analytically; and instead of laying down new names or new principles, and putting the science together, we begin from arithmetic, such as we know it, and leave all additional considerations till the want of them is felt. We shall thus see one new result spring up after another, until we find the necessity of speaking a new language, and giving interpretations to symbols which we did not at first contemplate. How this is done, and what it leads to, we cannot otherwise explain than by directing the student to proceed to the first chapter.

ELEMENTS OF ALGEBRA.

CHAPTER I.

EQUATIONS OF THE FIRST DEGREE.

We now proceed to the solution of equations *of the first degree.** This term must be explained.

To find the *degree* of a term generally, count the letters in it. Thus, abc is of the third degree; $aabc$ is of the fourth; for though there are only three letters, yet one of them occurs twice. The following are examples:

Of the first degree, a, b, c, x, z, p, &c.

Of the second degree, aa, bb, cx, bc, pz, &c.

Of the third degree, aaa, aab, abb, abc, pac, &c.

and so on.

To find the degree of a term with respect to any letters, count those letters only. Thus $3aaxxy$, of the fifth degree, is of the third degree with respect to x and y, of the fourth degree with respect to a and x, of the second degree with respect to x only, of the first degree with respect to y only, and so on. A term which does not contain x at all, is of *no degree* with respect to x, or is *independent* of x.

The degree of an equation with respect to any letter, is the degree of the highest term with respect to that letter. Thus the equation

$$xx - zxxx = yz - yyx$$

is of the third degree with respect to x, of the second with respect to y, and of the first with respect to z.

* Commonly called *simple* equations.

The solution of an equation of condition is the following problem:—Given an equation of condition, containing a letter the value of which is unknown; what is that number for which the unknown letter must stand, in order that the equation may be true? Are there more such numbers than one? if so, how many, and what are they? Or is there no such number, that is, is the equation impossible? and if so, how is that to be ascertained?

The scope of this will be better seen by some instances, which the pupil may verify.

The equation

$$2x - 1 = 5x - 19$$

is true then, and then only, when x is 6.

The equation

$$2x - 1 = 5x + 12$$

cannot be true, whatever x may stand for.

The equation

$$16x = 48 + xx$$

is true when x is 4, and is also true when x is 12; but never in any other case.

The equation

$$12x = 48 + xx$$

is never true for any value of x.

The equation

$$xxx + 11x = 6xx + 6$$

is true when x is 1, when x is 2, and when x is 3; and in no other case.

As an example of verification, let us try the latter equation when $x = 4$. Then

$$xxx = 64 \qquad 6xx = 96$$
$$11x = 44 \qquad 6xx + 6 = 102$$
$$xxx + 11x = 108$$

But 108 is not $= 102$; therefore $xxx + 11x$ is not $= 6xx + 6$, or $x = 4$ does not satisfy the equation.

We shall now use the following evident truths:

1. If equal numbers be added to equal numbers, the sums are equal numbers. That is, if $a = b$ and $c = d$, then $a + c = b + d$. If $a = b - c$ and $x = p - q$, then $a + x = (b - c) + (p - q) = b + p - c - q$. If $a = x - y$, and $b = x + y$, then $a + b = (x - y)$

$+(x+y) = x-y+x+y = 2x$. If $a = b+c$, then $a+v = b+c+v$.

2. If equal numbers be taken from equal numbers, the remainders are equal numbers. That is, if $a = b$ and $c = d$, then $a-c = b-d$. If $a = p-q$ and $b = p-2q$, then $a-b = (p-q)-(p-2q) = p-q-p+2q = q$. If $a = z+y$, then $a-m = z+y-m$.

3. If equal numbers be multiplied by equal numbers, the products are equal numbers. That is, if $a = b$ and $c = d$, $ac = bd$. If $a = b+c$, and $z = n$, then $az = n(b+c) = nb+nc$. If $d = l-v$, $2d = 2l-2v$.

If $$\frac{x}{2}+\frac{x}{3} = 1$$

Then $$2\frac{x}{2}+2\frac{x}{3} = 2 \text{ or } x+\frac{2x}{3} = 2$$

$$3x+3\frac{2x}{3} = 6$$

$$\text{or } 3x+2x = 6$$

4. If equal numbers be divided by equal numbers, the quotients are equal numbers. That is, if $a = b$ and $c = d$, then $\frac{a}{c} = \frac{b}{d}$. If $m = n$, then $\frac{m}{7} = \frac{n}{7}$. If $a = b-c$ and $p+q = z$, then $\frac{a}{p+q} = \frac{b-c}{z}$. If $7x = 14$, then $\frac{7x}{7} = \frac{14}{7}$, or $x = 2$.

The following abbreviations will be used, on account of the continual occurrence of the phrases:

$(+)a$ means, add a to both of the last-mentioned equal quantities, which gives ....

$(-)a$ .... subtract a from both of the last-mentioned equal quantities, which gives ....

$(\times)a$ .... multiply both the last-mentioned equal quantities by a, which gives ....

$(\div)a$ .... divide both the last-mentioned equal quantities by a, which gives ....

$(+)$ $(-)$ $(\times)$ and $(\div)$ by themselves I use to denote that the *two* last-mentioned sets of equal quantities are to be added. The following will explain the use of these abbreviations:

1.	$a = b-c$		$a-b = q+x$
$(+)c$	$a+c = b$	$(+)b$	$a = q+x+b$

2. $c-d = l-m$ $\qquad$ $2x-3 = 9$

$(+)\overline{d+m}$ $\quad c+m = l+d$ $\qquad$ $(+)3 \quad 2x = 12$

3. $p+q = a-b$ $\qquad$ $11x+18 = 100$

$(-)q$ $\quad p = a-b-q$ $\qquad$ $(-)18 \quad 11x = 82$

4. $p+q-z = 3a+4$

$(-)\overline{q-z}$ $\quad p = 3a+4-q+z$

5. $\frac{x}{2}-\frac{x}{3}+\frac{27}{4} = \frac{7x}{6}-\frac{5x}{12}+\frac{3}{4}$

$(\times)12$ $\quad \frac{12x}{2}-\frac{12x}{3}+\frac{324}{4} = \frac{84x}{6}-\frac{60x}{12}+\frac{36}{4}$

or $\quad 6x-4x+81 = 14x-5x+9$

6. $ax = b$ $\qquad$ $(a+b)x = c$

$(\div)a$ $\quad x = \frac{b}{a}$ $\qquad$ $(\div)\overline{a+b} \quad x = \frac{c}{a+b}$

7. $a-b+2c-3d = x-a+b$

$b+3c-2d = 4x-a-2b$

$(+)$ $\quad a+5c-5d = 5x-2a-b$

8. $2ax = b-z$

$a = b+z$

$(\div)$ $\quad 2x = \frac{b-z}{b+z}$

We shall now proceed to the solution of equations of the first degree, containing one unknown quantity, by means of the principles in pages 2 and 3, and the preceding operations.

1. What value of x will satisfy the equation

$3x-7 = x+19$

$(+)7$ $\quad 3x = x+26$

$(-)x$ $\quad 2x = 26$

$(\div)2$ $\quad x = 13$

Verification. If $x = 13$, $3x-\ 7 = 32$

$x+19 = 32$

2. $3x+16 = 10x+9$

$(-)3x$ $\quad 16 = 7x+9$

$(-)9$ $\quad 7 = 7x$

$(\div)7$ $\quad 1 = x$

Verification. If $x = 1,\ 3x+16 = 19$
$$10x+9 = 19$$

3. $$20x-13 = 102\tfrac{1}{2}-x$$

(+)13 $$20x = 115\tfrac{1}{2}-x$$

(+)x $$21x = 115\tfrac{1}{2} = \frac{231}{2}$$

(÷)21 $$x = \frac{231}{2\times 21} \quad \text{(Ar. 123.)}$$
$$= \frac{231}{42} = 5\tfrac{1}{2}$$

Verification. If $x = 5\tfrac{1}{2},\ 20x-13 = 97$
$$102\tfrac{1}{2}-x = 97$$

4. $$\frac{x}{2}+\frac{x}{3} = 1-\frac{x}{4}$$

(×)2 $$x+\frac{2x}{3} = 2-\frac{x}{2}$$

(×)2 $$2x+\frac{4x}{3} = 4-x$$

(×)3 $$6x+4x = 12-3x$$

(+)$3x$ $$6x+4x+3x = 12$$

that is $$13x = 12$$

(÷)13 $$x = \frac{12}{13}$$

Verification. If $x = \frac{12}{13}$, $\frac{x}{2}+\frac{x}{3}$ or $\frac{5}{6}x$ is $\frac{5}{6}$ of $\frac{12}{13}$, or $\frac{10}{13}$.
And $1-\frac{x}{4}$ is $1-\left(\frac{1}{4} \text{ of } \frac{12}{13}\right)$, or $1-\frac{3}{13}$, which is also $\frac{10}{13}$.

The same equation might be more easily solved by multiplying both sides by any common multiple of 2, 3, and 4. The least common multiple is the most advantageous; why, will appear on trying a higher one, as follows:

$$\frac{x}{2}+\frac{x}{3} = 1-\frac{x}{4}$$

36 is a common multiple of 2, 3, and 4.

(×)36 $$\frac{36x}{2}+\frac{36x}{3} = 36-\frac{36x}{4}$$

or $$18x+12x = 36-9x$$

(+)$9x$ $$18x+12x+9x = 36$$

that is $$39\,x = 36$$

(÷)39 $$x = \frac{36}{39},$$ which, reduced to its lowest terms, is $$= \frac{12}{13}$$

Now try 12, the *least* common multiple of 2, 3, and 4.

$$\frac{x}{2}+\frac{x}{3} = 1-\frac{x}{4}$$

(×)12 $$\frac{12\,x}{2}+\frac{12x}{3} = 12-\frac{12\,x}{4}$$

or $$6x+4x = 12-3\,x$$

Proceed as in the last case but one; and no reduction of the result to lower terms is necessary.

5. $$ab+a-b = 1$$

This equation differs from the preceding in having two unknown quantities. The real answer is, that there is an infinite number of values of a and b, which will satisfy this equation. If we *choose* a value of b, we can *find* the value of a, which, with the chosen value of b, will satisfy the equation. For instance, I ask, can b be $= 12$? Substitute 12 for b in the above equation, which then becomes

$$12a+a-12 = 1$$

or $$13a-12 = 1$$

(+)12 $$13\,a = 13$$

(÷)13 $$a = 1$$

The answer is, b may be 12 provided a be 1. In this case, $ab = 12$, and $12+1-12 = 1$.

Without making any *particular* assumption about the value of b, let us suppose it given; in what way must we combine this known b on the first side with the known unit on the second side, so as to point out the manner of finding a so soon as a particular value shall have been assigned to b?

Resume the equation:

$$ab+a-b = 1$$

(+)b $$ab+a = 1+b$$

But $ab+a$ is a taken one more than b times; that is, $ab+a = (1+b)\,a$.

Therefore $(1+b)a = 1+b$

$(\div)\ \overline{1+b}$ $a = \frac{1+b}{1+b} = 1$

The answer, then, is the following: b may be what we please, provided a be 1.

Verification. If $a = 1$,

$$ab+a-b = b+1-b = 1.$$

We have given this instance to shew how soon the operations of algebra lead to unexpected results. We will now take another instance.

6. $xy = x+y+1$

Knowing the value of y, to find that of x.

$(-)x$ $xy-x = y+1$

but $xy-x$ is x taken once less than y times, or $(y-1)x$. Therefore

$$(y-1)x = y+1$$

$(\div)\ \overline{y-1}$ $x = \frac{y+1}{y-1}$

Particular case. Let $y = 5$, then

$$x = \frac{5+1}{5-1} = \frac{6}{4} = \frac{3}{2}$$

Verification. $xy = \frac{3}{2}\times 5 = \frac{15}{2}$

$$x+y+1 = \frac{3}{2}+5+1 = \frac{3}{2}+\frac{10}{2}+\frac{2}{2} = \frac{15}{2}$$

General Verification. $x = \frac{y+1}{y-1}$ $xy = \frac{y(y+1)}{(y-1)}$

$$x+y+1 = \frac{y+1}{y-1}+y+1$$

$$= \frac{y+1}{y-1}+\frac{y(y-1)}{y-1}+\frac{y-1}{y-1}$$

$$= \frac{(y+1)+y(y-1)+(y-1)}{y-1}$$

$$= \frac{y+1+yy-y+y-1}{y-1}$$

$$= \frac{yy+y}{y-1} = \frac{y(y+1)}{y-1}$$

7. Two labourers can separately mow a field in 4 days and 7 days. They begin to work, and on the second day are joined by a third, who alone could mow the field in 10 days. The third remains

with the former two for a certain time, after which he leaves them; and it is then found that exactly four-fifths of the field have been mowed. How many days is this altogether?

[This question is introduced to shew how very soon algebraical symbols may be made to simplify complicated arithmetical reasoning.]

The fractions of the field which the first and second could mow in a day are $\frac{1}{4}$ and $\frac{1}{7}$. Let x be the whole number of days; or, that number being unknown, let x stand for it until it is known. Then the first man, who does one-fourth in one day, two-fourths in two days, &c., will in x days do x-fourths, or the fraction $\frac{x}{4}$ of the field. In the same time the second does $\frac{x}{7}$; but the third, who works one day less, at the rate of one-tenth a day, does $\frac{x-1}{10}$. Therefore, all that is done of the field is

$$\frac{x}{4}+\frac{x}{7}+\frac{x-1}{10}$$

But by the question this is $\frac{4}{5}$, which gives

$$\frac{x}{4}+\frac{x}{7}+\frac{x-1}{10} = \frac{4}{5}$$

The least common multiple of 4, 7, 10, and 5, is 140 (Ar. 103).

$(\times)140 \qquad \frac{140}{4}x+\frac{140}{7}x+\frac{140}{10}(x-1) = \frac{4\times 140}{5}$

or $\qquad 35x+20x+14(x-1) = 112$

or $\qquad 35x+20x+14x-14 = 112$

[because $\qquad 14(x-1) = 14x-14$]

therefore $\qquad 69x-14 = 112$

$(+)14 \qquad 69x = 126$

$(\div)69 \qquad x = \frac{126}{69} = \frac{42}{23} = 1\frac{19}{23}$

Verification.

In 1 day and $\frac{19}{23}$ of a day, there is mowed by the first $\frac{1}{4}$ and $\frac{19}{23}$ of $\frac{1}{4}$, or $\frac{42}{23}$ of $\frac{1}{4}$, or $\frac{21}{46}$ of the field; the second mows $\frac{42}{23}$ of $\frac{1}{7}$ or $\frac{6}{23}$, that is $\frac{12}{46}$; and the third, who works one day less than the others, or only $\frac{19}{23}$ of a day, does in that time $\frac{19}{23}$ of $\frac{1}{10}$, or $\frac{19}{230}$, that is $\frac{38}{460}$ of the field. But

$$\frac{21}{46}+\frac{12}{46}+\frac{38}{460} = \frac{368}{460} = \frac{4}{5}, \text{ as required.}$$

8. $$\frac{x-3}{2\frac{1}{2}}-\frac{x-4}{6\frac{1}{3}} = 3-\frac{x+1}{5}$$

A common multiple of $2\frac{1}{2}$ and $6\frac{1}{3}$ (not the least, which is 95, and with which the student should also solve the equation) is 570, which contains the first 228, and the second 90 times. It is also a multiple of 5.

$$(\times)570 \quad \frac{570}{2\frac{1}{2}}(x-3)-\frac{570}{6\frac{1}{3}}(x-4) = 1710-\frac{570}{5}(x+1)$$

or $$228(x-3)-90(x-4) = 1710-114(x+1)$$

But $$228(x-3) = 228x-684, \text{\&c.}$$

Therefore

$$(228x-684)-(90x-360) = 1710-(114x+114)$$

or $$228x-684-90x+360 = 1710-114x-114$$

$$(+)\ \overline{684+114x-360} \qquad 228x-90x+114x$$

$$= 1710-114+684-360 \text{ or } 252x = 1920$$

$$(\div)12 \qquad 21x = 160$$

$$(\div)21 \qquad x = \frac{160}{21} = 7\frac{13}{21}$$

Verification. If $$x = \frac{160}{21}$$

$$\frac{x-3}{2\frac{1}{2}} = \frac{\frac{97}{21}}{2\frac{1}{2}} = \frac{\frac{97}{21}}{\frac{5}{2}} = \frac{194}{105} = \frac{194}{5\times 21}$$

$$\frac{x-4}{6\frac{1}{3}} = \frac{\frac{76}{21}}{\frac{19}{3}} = \frac{228}{399} = \frac{228}{19\times 21}$$

$$\frac{x-3}{2\frac{1}{2}}-\frac{x-4}{6\frac{1}{3}} = \frac{194}{5\times 21}-\frac{228}{19\times 21} = \frac{2546}{5\times 19\times 21} = \frac{134}{5\times 21}$$

$$\frac{x+1}{5} = \frac{\frac{181}{21}}{5} = \frac{181}{5\times 21}$$

$$3-\frac{x+1}{5}=\frac{134}{5\times 21},$$ the same as before.

From these cases we may lay down the following rules for the solution of equations of the first degree.

1. *To clear an equation of fractions, multiply both sides by any common multiple of all the denominators: generally, the least common multiple is the most convenient.*

The following are some useful applications of this principle:

$$\frac{a}{b}=\frac{c}{d}\quad (\times)bd\quad \frac{abd}{b}=\frac{cbd}{d},\ \text{or}\ ad=bc$$

From the preceding equation the student is left to deduce the following:

$$a=\frac{cb}{d}\quad b=\frac{ad}{c}\quad c=\frac{ad}{b}\quad d=\frac{bc}{a}$$

$$\frac{1}{a}=\frac{d}{cb}\quad \frac{1}{b}=\frac{c}{ad}\quad \frac{1}{c}=\frac{b}{ad}\quad \frac{1}{d}=\frac{a}{bc}$$

From the first of the following equations let the student deduce all the rest.

$$\frac{ab}{xy}=\frac{cd}{pq}\quad a=\frac{cdxy}{pqb}\quad x=\frac{abpq}{cdy}\quad \frac{a}{y}=\frac{cdx}{pqb}$$

$$\frac{ab}{c}=\frac{dxy}{pq}\quad abpq=xycd\quad \frac{b}{dy}=\frac{cx}{apq}$$

2. *Any term of an equation may be removed from one side to the other if its sign be changed.* If this have not already occurred to the student from the preceding examples, it may be established by the following:

Let $\quad a+b=c+d-e$

$(-)b \quad a=c+d-e-b$

$(+)e \quad a+e=c+d-b$

In applying the rule for clearing an equation of fractions, care must be taken, when the denominator is removed, to remember that the sign which was placed before the complete fraction now belongs to the complete numerator, which should, therefore, be placed in brackets, or the proper rule for addition or subtraction applied at once. The following example will shew what is meant.

$$x+\frac{x-a}{b}-\frac{c+x}{ab}=d-\frac{x-e}{a}$$

$(\times)ab \quad abx+a(x-a)-(c+x)=abd-b(x-e)$

or $\quad abx+(ax-aa)-(c+x)=abd-(bx-be)$

or $\quad abx+ax-aa-c-x=abd-bx+be$

The mistake to which the beginner is liable is, to write $-c+x$ and $-bx-be$, instead of $-c-x$ and $-bx+be$.

By the second of the preceding rules,

$$abx + ax + bx - x = abd + aa + c + be$$

or $$(ab+a+b-1)x = abd+aa+c+be$$

$$(\div)\ \overline{ab+a+b-1} \qquad x = \frac{abd+aa+c+be}{ab+a+b-1} \quad \ldots .(1)$$

Verification.

$$x-a = \frac{abd+aa+c+be}{ab+a+b-1} - a$$

$$= \frac{abd+aa+c+be-a(ab+a+b-1)}{ab+a+b-1}$$

$$= \frac{abd+aa+c+be-aab-aa-ab+a}{ab+a+b-1}$$

$$= \frac{abd+a-ab-aab+c+be}{ab+a+b-1}$$

$$\frac{x-a}{b} = \frac{abd+a-ab-aab+c+be}{b(ab+a+b-1)} \quad \ldots\ldots (2)$$

And by similar processes,

$$\frac{c+x}{ab} = \frac{abc+ac+bc+abd+aa+be}{ab(ab+a+b-1)} \quad \ldots\ldots (3)$$

$$\frac{x-e}{a} = \frac{abd+aa+c-abe-ae+e}{a(ab+a+b-1)} \quad \ldots\ldots\ldots (4)$$

Reduce (1) (2) (3) and (4) to a common denominator, which can be done by multiplying the numerator and denominator of (1) by ab, of (2) by a, and of (4) by b. Then form

$$x + \frac{x-a}{b} - \frac{c+x}{ab}, \text{ or } (1)+(2)-(3)$$

which will be found to be

$$\frac{aabbd+aabd+abbe+abe-aab-abd-bc-be}{ab(ab+a+b-1)}$$

or $$\frac{aabd+aad+abe+ae-aa-ad-c-e}{a(ab+a+b-1)}$$

the latter of which arises from dividing the numerator and denominator of the former by b.

By similar processes,

$$d - \frac{x-e}{a} \text{ or } d-(4)$$

will be found to have the same value as $(1)+(2)-(3)$.

The student should not pass the preceding solution until he is able to repeat the whole on paper without the assistance of the book.

In the preceding equation, let c and e be each equal to nothing, which reduces the equation to

$$x + \frac{x-a}{b} - \frac{x}{ab} = d - \frac{x}{a};$$

the value of x to

$$x = \frac{abd + aa}{ab + a + b - 1};$$

and the value of each side of the equation, as just found, to

$$\frac{aabd + aad - aa - ad}{a(ab + a + b - 1)}$$

or $\frac{abd + ad - a - d}{ab + a + b - 1}$ (dividing numerator and denominator by a.)

These the student should find for himself from the equation.

We now look into particular problems to see what explanations may be necessary. Various unforeseen cases will present themselves; and each case will be explained, and a problem will be given for each, to shew how it may arise.

Anomaly 1. Let $a = 2 \quad b = 3 \quad d = \frac{1}{6}$

The equation then becomes

$$x + \frac{x-2}{3} - \frac{x}{6} = \frac{1}{6} - \frac{x}{2}$$

Then $x = \frac{2 \times 3 \times \frac{1}{6} + 2 \times 2}{2 \times 3 + 2 + 3 - 1} = \frac{5}{10} = \frac{1}{2}$

On attempting to verify the equation, we see that a contradiction appears; for x is $\frac{1}{2}$, and the operation $x - 2$, which is therefore impossible, appears in the second term. It seems, then, that *an equation may give a rational solution; and on attempting to verify the equation by this solution, the latter may be found to be impossible.* The question now is, can such an equation arise from a problem? if so, is it the problem itself which is absurd, or the way of treating it? If the latter, how is the method of solution to be set right?

Problem in illustration. A enters into this bargain with B, that he is to take B's property and pay his debts, taking his chance of gain or loss. On examination, it is found that B's property is (debts allowed for) exactly the same as that of A, with this exception only, that B is in partnership with another, and he and his partner have made a similar bargain with C. On examining C's affairs, he is found to be insolvent by £100. The result of the whole is, that this transaction falls short by £75 of making A's property twice as great as it was. What was A's property?

Let x stand for A's original property, which is, therefore, that of B and Co., independent of their share in the engagement with C. From the concluding paragraph we might presume that A is benefited by the transaction, namely, that the £x of B and his partner is more than sufficient to cover the loss arising out of their engagement with C. Let this be so; then $x - 100$ pounds remains for B and his partner, of which B's share, namely, $\frac{1}{2}(x - 100)$ is by the bargain transferred to A, who has, therefore,

$$x + \frac{x - 100}{2}$$

This doubles his property all but £75, and is, therefore, the same thing as $2x - 75$. Therefore,

$$x + \frac{x-100}{2} = 2x - 75$$

$$(\times)2 \qquad 2x + x - 100 = 4x - 150$$

$$4x - 2x - x = 150 - 100$$

$$x = 50$$

This introduces into the equation the anomaly we are now considering, for $\frac{50 - 100}{2}$ is impossible. It also contradicts the supposition on which it was obtained, namely, that x is greater than 100. We cannot, therefore, depend upon this solution. Suppose we try the other supposition, namely, that x is not greater than 100. In that case, B and his partner have to pay £100, of which they can only make good £x; of the remainder, or $100 - x$, A must, by the bargain, make good B's part, or $\frac{1}{2}(100 - x)$. This he loses by the transaction; and having x at first, he has now only

$$x - \frac{100 - x}{2}$$

This doubles his property all but £75, or rather we must now change this mode of speaking, which may seem to make one part of the problem disagree with another, and say simply that A's property is £75 less than twice what he had before. Hence,

$$x - \frac{100 - x}{2} = 2x - 75$$

$$(\times)2 \qquad 2x - (100 - x) = 4x - 150$$

or $$2x-100+x = 4x-150$$

or $$2x+x-100 = 4x-150$$

which is now the same equation as before, and not absurd in its present form; for though (since it yields $x = 50$) $x - 100$ is absurd, yet $2x + x - 100$ is not so. Hence we see that the effects of the wrong supposition, which made us write $x + \frac{1}{2}(x - 100)$, where we should have written $x - \frac{1}{2}(100 - x)$, disappear in resolving the equation, and leave the same result as we should have obtained by proceeding correctly.

The anomaly arises from an error of this sort. If b be greater than c, we know that

$$a-(b-c) = a-b+c$$

but if we have $a - b + c$, and wish to bracket b and c together, we cannot do this correctly until we know which is the greater. If it be b, the preceding is $a-(b-c)$;

but if it be c, this should be

$$a+(c-b).$$

One or other of the preceding two is absurd, except only when $b = c$, which makes them $a - 0$ and $a + 0$, or a for both.

From hence we may see, so far as one instance can shew it, that any mistake which amounts to no more than writing $a - (b - c)$ instead of $a + (c - b)$, as the representative of $a - b + c$, makes no difference in the final result. We here write, side by side, the solution of two equations, which only differ as above.

$x-\frac{a-x}{b} = c+\frac{x-b}{b}$	$x+\frac{x-a}{b} = c-\frac{b-x}{b}$
$bx-(a-x) = bc+(x-b)$	$bx+(x-a) = bc-(b-x)$
$bx-a+x = bc+x-b$	$bx+x-a = bc-b+x$

The remaining part is common to both.

$$bx = bc+a-b$$

$$x = \frac{bc+a-b}{b}$$

Anomaly (2). Let

$$ax+b = cx+d$$

$$ax-cx = d-b$$

$$(a-c)x = d-b$$

$$x = \frac{d-b}{a-c}$$

Let it happen that d is less than b, but a greater than c, as in the case of

$$3x + 4 = 2x + 1$$

There is then an impossible subtraction in the numerator of the result; and it is sufficiently evident, on other grounds, that the equation is impossible; for if a be greater than c, ax must be greater than cx; from which, if b be greater than d, $ax + b$ must be greater than $cx + d$, and cannot be equal to it. A similar question may now be proposed to that which arose upon the last anomaly. (See p. 12.)

PROBLEM I. IN ILLUSTRATION. In the year 1830, A's age was 50 and B's 35. Give the date at which A is twice as old as B.

This must either be before or after 1830. Try the second case, and let the required date be $1830 + x$.

Then A's age *will be* $\quad 50 + x$

.. B's .... $\quad 35 + x$

and
$$50 + x = 2(35 + x)$$
or
$$50 + x = 70 + 2x$$
$$2x - x = 50 - 70$$

Here we see an impossible subtraction, and it is also evident that $2x + 70$ must be greater than $x + 50$. Now, try a date *before* 1830; say $1830 - x$.

Then A's age *was* $\quad 50 - x$

.. B's .... $\quad 35 - x$

and
$$50 - x = 2(35 - x)$$
or
$$50 - x = 70 - 2x$$
$$2x - x = 70 - 50$$
or
$$x = 20$$

An evidently true answer; for in the year $1830 - 20$, or 1810, A's age was 30 and B's was 15. Here then we see that an impossible subtraction may arise from assuming a date to be after a certain epoch which is in fact before it, or *vice versâ*.

PROBLEM II. A and B have accounts together. The state of their affairs is this: Give A half as much as will make their dealings worth £500 to him, and give B £100, and they will then, after settling their account, have equal sums. How does their account stand?

The balance is either in B's favour or in A's. Take the latter, and suppose A ought to receive £x. Then $500 - x$ will make this transaction worth £500 to him, because

$$x + (500 - x) = 500$$

Give him $\frac{1}{2}$ (500 — x) and he will have (when B pays him)

$$x + \frac{500 - x}{2}$$

Now B, when he gets the £100, must pay £x to A, and will therefore have £(100—x). But they have then equal sums; therefore,

$$x + \frac{500 - x}{2} = 100 - x$$

$$(\times)2 \quad 2x + (500 - x) = 200 - 2x$$

or

$$2x + 500 - x = 200 - 2x$$

$$2x + 2x - x = 200 - 500$$

or

$$3x = 200 - 500$$

which is impossible. Try the other case, and suppose the balance to be in B's favour, and that he ought to receive £x. Then, to make A worth £500, his debt to B must be paid, and he must receive £500 besides: that is, he must receive 500 + x. But of this one-half only is given, or $\frac{1}{2}$ (500 + x), out of which, when he pays £x to B, he will have

$$\frac{500 + x}{2} - x$$

Now B gets £100 and also £x from A, and will therefore have 100 + x pounds. And, since they have then equal sums,

$$\frac{500 + x}{2} - x = 100 + x$$

$$(\times)2 \quad 500 + x - 2x = 200 + 2x$$

$$2x + 2x - x = 500 - 200$$

or

$$3x = 300 \text{ and } x = 100$$

Therefore A owes B £100.

Problem III. A traveller proceeds along a road on which, at various intervals, are found direction-posts variously numbered, pointing north or south. As soon as he reaches No. 1, he proceeds in the direction pointed out by it till he reaches No. 2, and so on. He finds the first direction-post after he has travelled 16 miles north, and he finds also that he changes his direction at every post which he meets *after the first*;* that the distance between every two posts is double

* The problem does not say whether he changes his direction at the first, or not.

that between the preceding, and that, at the fifth post, he is 86 miles north of his first position. What is the arrangement and character of the posts?

After travelling 16 miles north he reaches a post, and we are not told whether he is there directed to go on or turn back. Let us suppose the former, and that he travels x miles further north before he reaches the second post. He will then be $16+x$ miles north of his first position. At the next post he has to turn back, and proceed $2x$ miles south. Here two cases arise. If $2x$ be less than $16+x$, the third post will be north of his first position by $16+x-2x$ miles; but if $2x$ be greater than $16+x$, the third post will be south of his first position by $2x-(16+x)$ miles. Suppose the first; then between the third and fourth post there are $4x$ miles, and he has to go north from the third, therefore he will meet the fourth post at $16+x-2x+4x$ miles north of his first position. At the fourth post he turns south, and after proceeding $8x$ miles, meets the fifth post, his position north of his first position being then $16+x-2x+4x-8x$ miles. But this by the problem is 86 miles: consequently

$$16+x-2x+4x-8x = 86$$

or

$$8x-4x+2x-x = 16-86$$

or

$$5x = 16-86$$

in which there is an impossible subtraction. Let us now try the other hypothesis, and suppose that at the first direction-post he has to turn south, and that he finds the second after proceeding x miles south, or at $16-x$ north of his first position, or $x-16$ south, according as x is less or greater than 16. By the same attention to the expressed conditions of the problem, we find that

$$16-x+2x-4x+8x = 86$$

$$8x-4x+2x-x = 86-16$$

$$5x = 70$$

$$x = 14$$

Consequently the positions of the posts are as follows:

South ———— North

(4)☞		(2)☞	☜(1)	☜(3)	(5)
26	+ His first position.	2	16	30	86

Under each is marked the number of miles at which it is from his first position.

If we collect together and look at the correct and incorrect equations which we have found in the three preceding problems, we shall have the following:

PROBLEM I.

Incorrect, $50+x = 2(35+x)$ or $x = 50-70$ years *after* 1830

Correct, $50-x = 2(35-x)$ or $x = 70-50$ .. *before* ..

PROBLEM II.

Incorrect, $\frac{500-x}{2}+x = 100-x$ or $x = \frac{200-500}{3}$ which B owes A

Correct, $\frac{500+x}{2}-x = 100+x$ or $x = \frac{500-200}{3}$ .. A .. B

PROBLEM III.

Incorrect, $16+x-2x+4x-8x=86$ or $x=\frac{16-86}{5}$ miles *north*

Correct, $16-x+2x-4x+8x=86$ or $x=\frac{86-16}{5}$ miles *south*

From which, as well as from other instances, the following principle is clear:

When the value of x, deduced from an equation, contains an impossible subtraction, both the equation and the meaning of x have been misunderstood, and require alteration.

1. To correct the equation, alter the sign of every term which contains x *once only as a factor*.

[The words in italics are inserted to remind the student that we cannot draw any conclusion from the preceding as to equations which contain such terms as xx, xxx, &c. All our equations have been *of the first degree*.]

2. To correct the result, invert the terms of the impossible subtraction (that is, change 50—70 into 70—50), and let the quality of the answer be the direct reverse of that which was supposed when the incorrect equation was obtained. Thus, change years *after* into years *before*; *property* into *debt*; distance measured in one direction into that exactly opposite; and so on. Or, whatever alternatives it may be possible to choose between in assuming x, provided one be the direct reverse of the other, then, if one alternative produce an impossible subtraction in the value of x, the other is the one which should have been chosen.

An equation generally obliges us to take a more extensive view of the question than the words of the problem will bear, and will frequently shew that the view taken of the problem is not in every part a consistent whole.

In the preceding questions we have taken care to leave every possible case open in the statement of the problem: thus we have said (Problem I.), "Give the date at which A is twice as old as B," not "How long will it be before A is twice as old as B?" because the latter would be tacitly assuming that the event is to come, whereas it would be found out that the event is past, and the implied statement is erroneous. We write underneath the correct and incorrect mode of enunciating the question.

CORRECT.	INCORRECT.
In the year 1830, A's age was 50 and B's 35. *Give the date* at which A is twice as old as B.	In the year 1830, A's age was 50 and B's 35. When *will* A be twice as old as B?
ANSWER.	ANSWER.
20 years *before* 1830, or in 1810.	Never; but A *was* twice as old as B 20 years *ago*.

The chance of an impossible subtraction occurs in both; but in the first it arises from a question being left open to the student, who may choose the wrong alternative; in the second it arises from a wrong alternative being already tacitly assumed in the problem.

The alternatives presented by a problem may generally be ascertained with ease; but if not, the equation itself is frequently a guide.

Anomaly 3. Let

$$3x-10 = 2x-8$$
$$3x-2x = 10-8$$
$$x = 2$$

On verifying this equation we find each side to contain an impossible subtraction, for $3x-10$ is $6-10$, and $2x-8$ is $4-8$. After what has been said on the last case, we need not dwell upon this; the problem in the next page will furnish an instance. Somewhat similar to this is a mistake in the process which may introduce an impossible subtraction into both numerator and denominator of the answer. If, in solving $ax+b = cx+d$, thus, $ax-cx = d-b$ or $x = \frac{d-b}{a-c}$, we afterwards find a less than c, and

d less than b, it is a sign that we ought to have chosen the process $cx - ax = b - d$. This is a mistake in the order of operations only, and not in the conception of the problem.

PROBLEM. Divide the number 13 into two parts, in such a manner that three times the first may exceed half the second as much as the first exceeds 4.

It will be found from the resulting equation, namely,

$$3x - \frac{13-x}{2} = x-4$$

(where x is the first part) that $x = 1$, and therefore the parts of 13 required should be 1 and 12. But the problem is then impossible, for three times the first does not exceed half the second. But if the words "fall short of" be substituted for "exceed" throughout the problem, the equation becomes

$$\frac{13-x}{2} - 3x = 4-x$$

the answer to which is $x = 1$ as before, and the problem is possible; for $3x$ or 3 falls short of $\frac{1}{2}(13-x)$ or 6, as much as x or 1 falls short of 4,

We now ask how it happens than an equation gives a rational result, by which, when it is tried, the equation itself is proved to be irrational; and are we to conclude that no answer to an equation holds good until it has been tried upon the equation, and found to satisfy it rationally? Let us examine our first instance. The equation is

$$3x-10 = 2x-8$$

and the answer $x = 2$, when applied to the equation, gives

$$6-10 = 4-8$$

It so happens that the rules for solving an equation give the same answer to both of the following:

$$3x-10 = 2x-8$$
$$10-3x = 8-2x$$

And the following example will shew how this happens:

$$ax-b = cx-d \qquad\qquad b-ax = d-cx$$
$$ax-cx = b-d \qquad\qquad b-d = ax-cx$$
$$x = \frac{b-d}{a-c} \text{ in both.}$$

Hence, when an anomaly of the kind treated in this article occurs, it is the sign of a misconception of the right way of viewing the problem, but of a misconception which no way affects the result.

Anomaly 4. If we first solve the equation

$$ax+b = cx+d$$

which gives $x = \frac{d-b}{a-c}$, and if it should happen (without our observing it during the process) that $a = c$ or $a - c = 0$, we shall then find an answer of the following form,

$$x = \frac{d-b}{0}$$

which is unintelligible; because there can be no answer to the question, "How often is *nothing* contained in $d-b$?" or, at least, if there be any answer, it is, "Nothing, however often it may be repeated, yields nothing, and therefore cannot be repeated often enough to yield $d-b$." On returning to the equation, we see that the supposition of $a = c$ gives $ax = cx$, so that, as far as the equation only is concerned, it is always true if $b = d$, and never true if b is unequal to d. But in giving further explanation of problems which produce such equations, we shall employ the following principle, the propriety of which is obvious.

When any supposition (such for instance as making $a = c$ in the preceding equation), makes the results of ordinary rules unintelligible, then, instead of making a *exactly* equal to c, let it be made *very nearly* equal to c, and observe the result: afterwards suppose it still nearer to c, and so on; the succession of results will inform us whether any rational interpretation can be put upon the result of supposing $a = c$, or not. We shall now try a problem in which the preceding difficulty will be found to occur.

Problem. There are three trading companies, of 4000, 5000, and 9000 shares respectively, and all three would, if broken up, pay the same dividend upon their shares; but if every shareholder in the second advanced his company £10 on each share, and every shareholder of the third advanced £12 in like manner, then the first two companies together would have the same total assets as the third; what is the dividend which each company could now pay?

Let x pounds per share be that dividend. Then, after the advances supposed in the problem, the three companies could pay x, $x+10$, and $x+12$ pounds per share respectively; which, taking their number of shares into account, supposes them to be in possession of $4000x$,

5000 $(x+10)$, and 9000 $(x+12)$ pounds respectively: between which the last clause of the problem gives the equation

$$4000x + 5000(x+10) = 9000(x+12)$$

$$(\div)1000 \qquad 4x + 5(x+10) = 9(x+12)$$

or

$$9x + 50 = 9x + 108$$

In which we recognise the anomaly which is the subject of this article. Nor can we in this case, as in page 18, account for the impossibility by supposing that we have mistaken the problem, and that the three societies are at first insolvent, and are in debt upon each share; for, if we make this supposition, and let x be the amount they fail for upon each share, we have already seen (and the student must make himself sure of the same in this particular case) that the resulting equation will be

$$50 - 9x = 108 - 9x$$

which is equally impossible with the former. We shall, therefore, now try the consequences of a slight change in the conditions, agreeably to the preceding principle. For instance, we will suppose the third society to have only 8999 shares, instead of 9000. The equation then becomes

$$4000x + 5000(x+10) = 8999(x+12)$$

or

$$4000x + 5000x + 50{,}000 = 8999x + 107{,}988$$

$$4000x + 5000x - 8999x = 57{,}988$$

or

$$x = 57{,}988$$

The answer therefore is, that each society could at first pay £57,988 per share. Let us try the effect of a still smaller change, and suppose the third society to want only the hundredth part of a share of 9000 shares, that is, to have $8999\frac{99}{100}$ shares. Then we have

$$4000x + 5000(x+10) = 8999\frac{99}{100}(x+12)$$

$$4000x + 5000x + 50{,}000 = 8999\frac{99}{100}x + 8999\frac{99}{100} \times 12$$

$$4000x + 5000x - 8999\frac{99}{100}x = 8999\frac{99}{100} \times 12 - 50{,}000$$

$$\frac{1}{100}x = \frac{899999}{100} \times 12 - 50{,}000$$

$$(\times)100 \qquad x = 899{,}999 \times 12 - 5{,}000{,}000$$

$$x = 5{,}799{,}988$$

or each society can pay £5,799,988 per share. In the same way, if we take the third society at $8999\frac{999}{1000}$ shares, we shall yet have a still greater answer, and so on. A similar result would be obtained by increasing the 4000 or 5000 shares of either of the other societies. Therefore, the answer to the preceding question is, that no number is great enough to satisfy the conditions of the question; but that if these conditions be slightly altered, an answer may be found, which answer is a greater number the slighter the alteration just alluded to. Dismissing the problem, which we have only introduced to shew that such anomalies may arise in the application of algebra, we return to the consideration of similar equations.

The solution of

$$ax = bx + c$$

is

$$x = \frac{c}{a-b}$$

in which, should it happen that $a = b$, the answer is unintelligible, being $\frac{c}{0}$, and the equation impossible, being

$$ax = ax + c$$

but if a exceed b by any quantity, however small, the equation and its answer are both rational. Let a exceed b by the fraction of unity $\frac{1}{m}$, then the equation becomes

$$\left(b + \frac{1}{m}\right) x = bx + c$$

or

$$bx + \frac{x}{m} = bx + c$$

$$(-)bx \quad \frac{x}{m} = c \qquad (\times)m \quad x = mc$$

The same might be obtained from the preceding answer, for

$$x = \frac{c}{a-b} = \frac{c}{\frac{1}{m}} = mc$$

To make a exceed b by a small quantity, $\frac{1}{m}$ must be small, that is, m must be large; and in this way we may get an equation whose answer shall be as large as we please. For instance, let c be 1; and suppose we want an equation of the preceding form whose answer shall be 1,000,000. Let $m = \frac{1}{1{,}000{,}000}$. Then such equations as the following have the answer $x = 1{,}000{,}000$.

$$7\frac{1}{1{,}000{,}000}\,x = 7x + 1$$

A number which does not satisfy an equation may, we can easily conceive, *nearly* satisfy it. But this word *nearly* is too indefinite for our purpose, as we shall now shew. Suppose we have the equation $7x = 2x + 3$, and we try whether $x = 1$ will satisfy this equation. It will not; for the first side is 7 (x being 1), and the second side is 5; that is, the first side, instead of being equal to the second, is greater by 2. The same result applies, in the same words, to the equation $7x = 5x + 1988$, if we try $x = 1000$ upon it; for the first side becomes 7000, and the second 6988. Shall we then say, $x = 1$ satisfies the first as nearly as $x = 1000$ does the second? Is 7 as near to 5 as 7000 to 6988? If we look at the differences only, we must answer yes; for

$$7 - 5 = 2$$
$$7000 - 6998 = 2$$

but in the common use of the word near* (to which it will be convenient to keep) it would be said that 7000 is nearer to 6998 than 7 is to 5. In the first case, the difference is 2 out of 7000; in the second it is 2 out of 7. Keeping to this meaning of the term, we shall in future consider ax and $ax+c$ as nearer to equality when x is greater than when it is smaller. And in this sense we say, that when a problem leads to such an equation as

$$ax = ax + c$$

the result is, no number is great enough to be an answer to the problem; but the greater any number, the more nearly is it an answer to the problem.

Tried by the common rules (which, in this case, lead us too far) the answer to the preceding equation is

$$x = \frac{c}{a-a} \text{ or } \frac{c}{0}$$

* In the making of a bargain, £6998 would be considered as the same price, within a *trifle*, as £7000; but any thing at £7 would be considered dear as compared with the same at £5. We might multiply instances in which the same quantity would be considered small under some circumstances, and great under others.

and it is customary to say, that $\frac{c}{0}$ means an *infinite number*, and that the answer is *infinitely great*. Taken literally, such phrases are unmeaning, because we know of no number which is infinitely great, that is, greater than can be counted or measured. But the word is often used; and we shall, therefore, adopt it with the following meaning:

By saying that $\frac{1}{q}$ is infinite when $q = 0$, we are to be understood as meaning no more than a short way of expressing the following:—When q is small, $\frac{1}{q}$ is great; if q become still smaller, $\frac{1}{q}$ is still greater, and so on; so that $\frac{1}{q}$ may be made to exceed any given number, however great, if q be taken sufficiently small. And when we say the answer to a problem is infinite, we mean that no number is great enough to satisfy the conditions of the question; but that any great number nearly does so, a still greater still more nearly, and so on; so that the problem may be answered within any degree of nearness (short of positive exactness) by taking a sufficiently great number.

Anomaly 5. The solution of

$$ax + b = cx + d$$

is

$$x = \frac{d-b}{a-c}$$

Now, if it should happen that, after the solution of this equation, it becomes necessary to suppose a equal to c (as in the last case), and b also equal to d, the answer

$$\frac{d-b}{a-c} \text{ must be written } \frac{0}{0}$$

which has no meaning. On returning to the equation, we no longer find any necessity to give x one value rather than another; for, if a be $= c$, and $b = d$, then $ax + b$ is equal to $cx + d$, whatever x may be. Therefore the answer to the question is, that every possible value of x satisfies the conditions. We shall apply this in the following

PROBLEM. Is there any number such that a times one less than the number, added to b times two more than the number, is exactly c times the number; a, b, and c, being given numbers or fractions?

Let x be that number, then

$$a(x-1)+b(x+2) = cx$$
$$ax-a+bx+2b = cx$$
$$ax+bx-cx = a-2b$$
$$(a+b-c)x = a-2b$$
$$x = \frac{a-2b}{a+b-c}$$

Verification.

$$x-1 = \frac{c-3b}{a+b-c} \qquad a(x-1) = \frac{ac-3ab}{a+b-c}$$
$$x+2 = \frac{3a-2c}{a+b-c} \qquad b(x+2) = \frac{3ab-2bc}{a+b-c}$$
$$a(x-1)+b(x+2) = \frac{ac-2bc}{a+b-c} = \frac{c(a-2b)}{a+b-c}$$
$$= c \times \frac{a-2b}{a+b-c} = cx$$

Suppose that a is 8, b is 4, and c is 12, or that the problem is the following:—What number is that, one less than which multiplied by 8, added to the product of 2 more and 4, is equal to 12 times the number? Here we find $a-2b=0$, and $a+b-c=0$; so that the preceding answer takes the form $\frac{0}{0}$: trying this case by itself, to find out the reason, we have the equation,

$$8(x-1)+4(x+2) = 12x$$
$$8x-8+4x+8 = 12x$$
$$12x = 12x$$

which being always true, the answer is, that every number and fraction whatsoever, which is greater than 1, satisfies the conditions of the problem; a result corresponding to the interpretation we have already seen reason to put upon the form $\frac{0}{0}$.

We shall now proceed to the solution of some problems, in which the preceding method and principles will be applied. As instances will be taken from different parts of natural philosophy, we shall divide them into sections, and state at the head of each the facts on which the solution depends.

EXAMPLES.—Section I. *Specific gravities.*—By the specific gravity of a body is meant the number of times which its weight is of the weight of an equal bulk of water. Thus, when we say that the specific gravity of *brick* is 2, we mean that a mass, say a cubic foot of brick, weighs twice as much as a cubic foot of water.* A cubic foot of water weighs about 1000 ounces avoirdupois.†

PROBLEM I. One pint of water is added to three pints of milk (specific gravity 1·03): what is the specific gravity of the mixture?

If a pint of water weigh m ounces, then one pint of milk weighs $1{\cdot}03 \times m$ ounces, and the whole four pints of the mixture weighs $m + (1{\cdot}03m) \times 3$, or $m + 3{\cdot}09\,m$ ounces. But four pints of water weigh $4m$ ounces; therefore the specific gravity of the mixture (see the preceding definition) is

$$\frac{m + 3{\cdot}09\,m}{4m} \quad \text{or} \quad \frac{4{\cdot}09\,m}{4m} \quad \text{or} \quad \frac{4{\cdot}09}{4} \quad \text{or} \quad 1{\cdot}0225$$

N.B. Here is an instance of a quantity m introduced for convenience, and which disappears in the process. The question itself is too simple to require an equation.

PROBLEM II. A number of cubic feet (m) of a substance whose specific gravity is a, is mixed with n cubic feet of another, having the specific gravity b. What is the specific gravity of the mixture?

It is plain that the $m+n$ cubic feet of mixture will be as heavy as $ma+nb$ cubic feet of water. Therefore the specific gravity required is $\dfrac{ma+nb}{m+n}$.

EXERCISE. Try to shew that $\dfrac{ma+nb}{m+n}$ must lie between a and b.

PROBLEM III. How much of a specific gravity 2 must be mixed with 20 cubic feet of specific gravity 10, in order that the specific gravity of the mixture may be 5?

Let x be the number of cubic feet in that quantity. Then the whole $20+x$ cubic feet of mixture has the same weight as $20 \times 10 + x \times 2$, or $200 + 2x$ cubic feet of water. Therefore the specific gravity is $\dfrac{200+2x}{20+x}$, and

$$\frac{200+2x}{20+x} = 5 \quad \text{or} \quad 200+2x = 5\,(20+x) \quad \therefore\ x = 33\tfrac{1}{3}$$

Generalisation of the preceding. How much of a specific gravity

* Atmospheric air is often taken as the standard of gases. Water is about 800 times as heavy as air.

† Easy to recollect, and remarkably near the truth. Let the student deduce it from AR. *Art.* 217.

a must be mixed with m cubic feet of a specific gravity b, in order that the specific gravity of the mixture may be c?

Let x be the quantity required. Then x cubic feet of specific gravity a weigh as much as ax cubic feet of water, and m cubic feet of specific gravity b as much as bm cubic feet of water. Hence the whole $m+x$ cubic feet of mixture weigh as much as $bm+ax$ cubic feet of water. Hence, as in the particular instance above,

$$\frac{bm+ax}{m+x} = c \qquad bm+ax = c(m+x)$$

$$= cm+cx$$

$$\therefore \quad bm-cm = cx-ax \quad \text{or} \quad (b-c)m = (c-a)x$$

$$x = \frac{b-c}{c-a}.m$$

This is rational when b is greater than c, and c greater than a; that is, when $b-c$ and $c-a$ are possible. It is also rational when $b-c$ and $c-a$ are *both* impossible; since, in this case, the apparent irrationality arises from our having converted

$$bm+ax = cm+cx \text{ into } bm-cm = cx-ax$$

$$\text{instead of } cm-bm = ax-cx$$

and the rational answer is $x = \frac{c-b}{a-c}.m$. In this case a is greater than c, and c greater than b. That is, this problem is rational when c lies between a and b. If c do not lie between a and b, then a rational problem is formed, as in page 18, by supposing x the direct reverse of what it was last supposed to be; that is, by supposing the m cubic feet of specific gravity b to allow of the substance of specific gravity a being subtracted from it, or to be itself a mixture already containing that substance. That is, solve this problem: How much of a specific gravity a must be *taken from* m cubic feet of specific gravity b, so that the specific gravity of the remainder may be c? The answer will be found to be

$$x = \frac{c-b}{c-a}.m \quad \text{or} \quad x = \frac{b-c}{a-c}.m$$

according as c is greater than both a and b, or less than both. But here may arise another of the anomalies previously explained. Suppose, for instance, we ask how much of a specific gravity $a(=10)$ must be taken from $m(=20)$ cubic feet of a specific gravity $b(=6)$, in order that the specific gravity of the remainder may be $c(=12)$.

(Call this problem A.) Here, though the problem is evidently impossible, the answer will be rational, being

$$x = \frac{c-b}{c-a}.m = \frac{12-6}{12-10}.20 = \frac{6}{2}.20 = 60$$

and the impossibility is detected, not in the form of the answer, but on looking at the problem, with which the answer is inconsistent; for 60 cubic feet cannot be taken from 20. The equation from which this answer results is

$$\frac{120-10x}{20-x} = 12 \text{ or } 120-10x = 240-12x$$

in which, with the answer $x = 60$, the *anomaly* 3 explained in p. 19 occurs. On correcting this equation, as done in p. 20, it becomes

$$10x-120 = 12x-240 \text{ or } \frac{10x-120}{x-20} = 12$$

which is derived from the following problem:—From how much of specific gravity $a (= 10)$ must $m (= 20)$ cubic feet of specific gravity $b (= 6)$ be taken, in order that the specific gravity of the remainder may be $c (= 12)$? (Call this problem B.)

Whether the answer $x = 60$ is to be called possible or impossible depends upon the answer to the following question. Was problem B within our meaning or not when problem A was proposed? that is, did we mean to take the one of the two, A and B, which should turn out to be rational; stating A, because we supposed it, before examination, to be that one? or did we mean to confine ourselves within the limits of the literal meaning of A? Because, in the first case, the answer is, that we have chosen the wrong alternative, that the other should have been chosen, and that the answer is $x = 60$; in the second case, the answer is that the problem is impossible.*

PROBLEM IV.—The specific gravities of gold and silver are $19\frac{1}{4}$ and $10\frac{1}{2}$; a goldsmith offers a mass of $\frac{1}{4}$ of a cubic foot which

* I have here stated this purposely, because the matter of convention is more obvious in this problem than in that of page 16, where there is an even chance of our choosing the wrong alternative at first. The problem before us will appear strained, simply because the alternatives of north and south of a post come more frequently into practice than those of taking a known from an unknown, and an unknown from a known, mixture. I state this because some writers on algebra seem to imply, by making this sort of extension rest only on the most obvious and usual examples, that they wish the student to consider it as not conventional, but necessary.

he asserts to be gold, and which is found to weigh 260 pounds. Can it be all gold? if not, may it have been adulterated with silver? and, in that case, in what proportion were silver and gold mixed?

Since a cubic foot of water weighs 1000 ounces, and gold is $19\frac{1}{4}$ times as heavy as water, a cubic foot of gold weighs 19,250 ounces, and $\frac{1}{4}$ of a cubic foot weighs $4812\frac{1}{2}$ ounces, or 300 pounds $12\frac{1}{2}$ ounces. Therefore the mass cannot be all gold. Again, a cubic foot of silver weighs 10,500 ounces, and $\frac{1}{4}$ of a cubic foot weighs 164 pounds 1 ounce. Therefore the mass is heavier than its bulk of silver, and lighter than its bulk of gold, and consequently may be a mixture of the two. In this case, let x be the quantity of gold, x being a fraction of a cubic foot; then $\frac{1}{4}-x$ is the remainder of silver, and $19250\,x$ is the number of ounces the gold weighs, and $10500\,(\frac{1}{4}-x)$ the same for the silver. But the whole weight is 260 pounds, or 4160 ounces; therefore

$$19250x + 10500(\tfrac{1}{4}-x) = 4160$$

or $$x = \frac{1535}{8750} = \frac{307}{1750} = \frac{3}{17} \text{ nearly, or } \frac{12}{17} \text{ of } \frac{1}{4}$$

∴ about 12 parts out of 17 are gold, and the rest silver.

This is a case of the celebrated problem first solved by Archimedes,* and may be generalised as follows:—In what proportions must substances of specific gravities a and b be mixed, so that the specific gravity of the whole may be c? To one cubic foot of the first let there be x cubic feet of the second, to produce the mixture required. Then the 1 cubic foot of the first, weighing a cubic feet of water, and x cubic feet of the second, weighing bx cubic feet of water, the whole $1+x$ cubic feet weigh $a+bx$ cubic feet of water. But since its specific gravity is to be c, it weighs $c(1+x)$ cubic feet of water; therefore,

$$a+bx = c(1+x), \quad \text{or} \quad x = \frac{a-c}{c-b}$$

to which the remarks in pages 28 and 29 apply.

EXAMPLES. Section II. *The Lever.* Generally speaking, a bar, when suspended by any point in it, will only rest in one position, namely, hanging downwards; but if properly loaded with weights, it may be made to rest in any position, and is then said to be a *lever* in *equilibrium*, that is, evenly balanced. The rules are, 1. Treat the

* See the *Penny Cyclopædia*, vol. ii. p. 277.

weight of the bar itself as if it were all collected in its midddle point. 2. Call the number of *pounds** in any weight, multiplied by the number of *feet* by which it is removed from the point of suspension, the *moment* of that weight; then a bar will be in equilibrium when the sum of the moments of the weights on one side of the pivot is the same as that of the weights on the other side. If the bar be not suspended by its middle point, the weight of the bar itself must be taken into account as if it were all collected in the middle point. If the sum of the moments on one side be not equal to that on the other, the side which has the greater sum will preponderate.

PROBLEM I. A bar 18 feet long, weighing 40 pounds, has weights of 12 and 20 pounds at the two ends. Where must the pivot be placed, so that the bar may rest upon it?

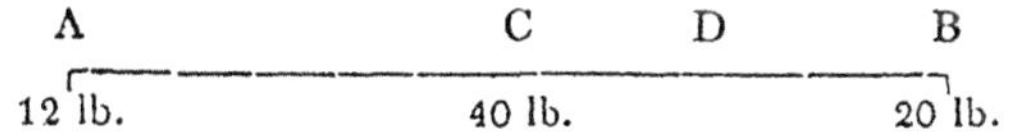

Let the weight of the bar (40 lbs.) be collected at the middle point C. Then $AC = CB = 9$ feet. We do not know on which side of C to place the pivot, which may produce an incorrectness in the equation similar to that in *Anomaly* 1. page 12. This, however, will not affect the result. Let the pivot be at D between B and C; that is, let 12 lb. at A and 40 lb. at C balance 20 lb. at B. Let $AD = x$ feet. Then $CD = (x-9)$ feet, $DB = (18-x)$ feet. The *moments* of the weights are $12x$, $40(x-9)$ and $20(18-x)$; and by the preceding principle,

$$12x + 40(x-9) = 20(18-x), \quad \text{or} \quad x = 10;$$

therefore the pivot is 1 foot to the right of the middle point, and the problem has been rightly interpreted, because $x-9$ and $18-x$, tried by the result, are both possible. If we had imagined D to be on the left of C, and $AD = x$ as before, we should have supposed the weights at C and B conspiring to balance that at A, and $DB = 18-x$ as before, but $CD = 9-x$ instead of $x-9$. The equation would have been

$$12x = 40(9-x) + 20(18-x)$$

$$\text{or} \quad 12x - 40(9-x) = 20(18-x) \quad \text{or} \quad x = 10$$

* Any other units may be substituted for pounds and feet; but care must be taken to use the same units throughout the whole of each problem.

differing from the first in having

$$-40\,(9-x) \text{ instead of } +40\,(x-9).$$

The solution $x=10$, would have shewn us the *Anomaly* 1, page 12.

To generalise this problem, let the length of the bar be l, its weight* W, the weights at the left and right extremities P and Q. Then let $AD=x$, which gives (supposing the pivot right of C), $CD=x-\frac{1}{2}l$, $DB=l-x$. The equation becomes

$$Px+W(x-\tfrac{1}{2}l) = Q\,(l-x)$$

whence $$x = \frac{Wl+2Ql}{2P+2Q+2W} = \frac{W+2Q}{P+Q+W} \times \frac{l}{2}$$

Exercise. Prove from the value of x just found that x is greater than, equal to, or less than, $\frac{1}{2}l$, according as Q is greater than, equal to, or less than P.

The preceding problem contains the principle of the steelyard.

Problem II. If a bar 20 feet long, weighing 6 pounds, be supported at 9 feet from the left extremity, how must we place upon it weights of 16 and 17 pounds, 7 feet apart, so that the whole may be balanced (the 16 lbs. being supposed on the left)?

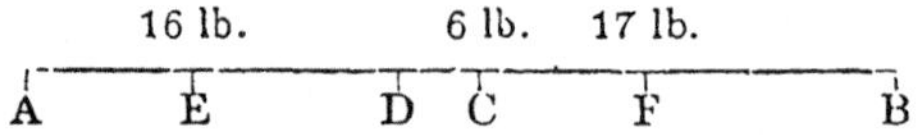

Let C be the middle point, D the pivot, and E and F the places of the weights. Then $AD=9$ feet, $BD=11$ feet, $AC=10$ feet, $EF=7$ feet, $DC=1$ foot. Let $AE=x$; then $ED=9-x$, $DF=AF-AD=AE+EF-AD=x+7-9=x-2$. The system therefore consists in

16 lb. at dist. $\overline{9-x}$ from pivot; moment $16\,(9-x)$,

which balances

6 lb. at dist. 1 foot from pivot; moment 6×1 or 6

17 lb. at dist. $\overline{x-2}$ feet from pivot; moment $17\,(x-2)$.

Hence $$16\,(9-x) = 6+17\,(x-2) \qquad x = 5\tfrac{7}{33} \text{ feet.}$$

The equation has been rightly formed, for $9-x$ and $x-2$ are both possible. If we had supposed E and F to fall on the same side of D, the resulting equation would have been

* That is, W is the number of pounds, or other unit, in its weight, and l the number of feet, or other unit, in its length.

$$16(9-x)+17(2-x) = 6$$

or $$16(9-x) = 6-17(2-x) \qquad x = 5\tfrac{7}{33} \text{ feet:}$$

from which the same value of x is obtained, but $2-x$ is impossible. Even if we had made the evidently impossible supposition that E and F are both on the same side of D as C (amounting to supposing that the weights at E, F, and C are counterbalanced by no weight at all on the opposite side), the equation which results, treated by ordinary rules, would present no direct sign of impossibility until we came to compare the result with the equation. In this case,

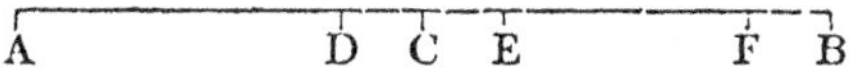

if $AE=x$, we have $DC=1$, $DE=x-9$, $DF=(x-9)+7=x-2$; and the moments of the weights at C, E, and F, are therefore 6, $16(x-9)$ and $17(x-2)$. To trace the effect of the preceding supposition, namely, that there is no weight on the left side of the pivot D, we must see what will follow from supposing

$$6+16(x-9)+17(x-2) = 0$$

as if such an equation were possible. This gives

$$6+16x-144+17x-34 = 0$$

$$33x-172 = 0 \qquad 33x = 172 \qquad x = 5\tfrac{7}{33}$$

the same as before. But the preceding equation is impossible, since the addition of three quantities must give more than nothing. At the same time, we see that $x-9$ is also impossible.

We must here make a remark similar to that in page 14. The equation

$$x-(c-b) = 0, \quad \text{or} \quad x+b-c = 0$$

is possible, since it merely indicates that $x=c-b$. But the equation

$$x+(b-c) = 0$$

is impossible. Nevertheless, when b is greater than c, $x+(b-c)$ is the same thing as $x+b-c$; and by attempting this conversion when b is less than c, we might be led to the impossible form

$$x+(b-c) = 0,$$

where we should have adopted the rational form

$$x-(c-b) = 0.$$

From this we conclude, that if we meet with

$$x+p = 0,$$

it is a sign that, in forming the quantity p, we have inverted the order of the terms in a subtraction; that is, we have supposed p was $b-c$ when it should have been $c-b$. Let us call the latter q; then if we had proceeded correctly, we should have had

$$x-q=0, \quad \text{or} \quad x=q.$$

The problem in page 32, generalised, is as follows. A bar l feet in length, weighing W pounds, is supported at a feet from its left extremity. How must we place P and Q pounds (P being on the left) m feet asunder, so that the bar may be balanced? The equation is

$$P(a-x)=W(\tfrac{1}{2}l-a)+Q(x+m-a)$$

$$x=\frac{Pa+Qa+Wa-\frac{1}{2}Wl-Qm}{P+Q}.$$

Exercise. Supposing the bar to be continued to the left of A and the right of B, *but the continuation on either side to have no weight*, explain, as in page 33, the case where $W=20$ pounds, $l=50$ feet, $a=5$ feet, $P=4$ pounds, $Q=7$ pounds, $m=10$ feet.

Examples. Section III. *Miscellaneous*. Problem I.—A straight line AB, 10 inches long (continued both ways), is cut by the point C at 7 inches to the right of A. Where must the point D be placed, so that AC may bear the same proportion* to CB which AD bears to DB?

Here AC = 7 inches, CB = 3 inches. The point D must be either between A and B, or to the right of B, or to the left of A. Or there may be (for any thing we have shewn to the contrary), more than one such point; for instance, one such point to the right of A, and another to the left. But if we consider the conditions of the problem, it will appear impossible that D can lie any where but to the right of B. For, suppose D to be placed between A and B, say between C and B; then, according to the problem, AD (greater than 7 inches) contains BD (less than 3 inches), the same times and parts of a time which AC (7 inches) contains BC (3 inches). This, the

* To those who have never used this term mathematically (see Ar. 177, 178), we may state, that a bears to b the same proportion which c bears to d, when the fraction $\frac{a}{b}$ is the same as $\frac{c}{d}$.

least consideration will shew, cannot be. For a similar reason (this let the student explain), D cannot lie between A and C. Neither can D lie on the left of A; for then, of AD and DB, AD the less must contain DB the greater, as AC (7 inches) contains CB (3 inches), or $2\frac{1}{3}$ times, which is also impossible. Let us then suppose D on the right of B, and let $AD = x$ inches; then $BD = x - 10$ inches. By the problem, x bears to $x - 10$ the same proportion as 7 bears to 3; that is,

$$\frac{x}{x-10} = \frac{7}{3} \qquad (\times)3(x-10) \qquad 3x = 7(x-10)$$

which gives $x = \frac{35}{2} = 17\frac{1}{2}$ inches.

If we had supposed the point D to lie between A and B (say between C and B), then AD being x, DB would have been $10 - x$, and the equation would have been

$$\frac{x}{10-x} = \frac{7}{3}, \text{ which gives } x = 7;$$

that is, D coincides with C. This is a case which we have not put among what we have called anomalies, because the result, though not expected, is intelligible without further explanation. It implies, that if we would place D between A and B, so that AD should be to DB in the same proportion as AC to CB, D must occupy the same place as C. But if we suppose D to lie on the left of A, and let x stand for AD, we have $DB = 10 + x$, and the equation becomes

$$\frac{x}{10+x} = \frac{7}{3}, \quad \text{or} \quad 3x - 7x = 70,$$

which presents the anomaly in pages 14..19. This shews that we have measured x or AD in the wrong direction, and that if it had been made to fall to the right of A, the equation would have been

$$7x - 3x = 70, \quad \text{or} \quad x = 17\tfrac{1}{2} \text{ inches},$$

which we have found before.

Now let the problem be altered so that C shall stand in the middle between A and B. For instance, let $AB = 12$ inches, and $AC = 6$ inches. Is there a point D, not coinciding with C, such that AD bears to DB the same proportion as AC to CB? Certainly not; for AC contains CB once exactly, but if D be placed right of B or left of A, AD is greater or less than DB. But if a point be placed at a great distance from A or B on either side, AD contains DB very nearly once (see p. 24); and by removing D to a sufficient distance, AD may be made to contain D as nearly* once as we please. We might therefore expect, if we attempt to find D in this case by an equation, such an anomaly as that in page 21, explained in page 24. If possible, let the point D, on the right of B, satisfy the conditions of this problem. Let AD $=x$. Then DB $=x-12$ and AC $=6$, CB $=6$. Therefore the equation of the problem is

$$\frac{x}{x-12}=\frac{6}{6}=1 \quad \text{or} \quad x=x-12 \quad \text{(See page 21.)}$$

The generalisation of the preceding problem, in supposing AB $=a$, AC $=b$, and AD $=x$, and placing D on the right of B, gives the equation

$$\frac{x}{x-a}=\frac{b}{a-b} \quad \text{or} \quad x=\frac{ab}{2b-a}$$

In order to meet all the cases which may occur in the application of algebra, we will now take a problem in which the answer will go beyond the notion which was formed when the problem was proposed, not because the thing proposed to be done is impossible, but because the answer is not within the limits of what is usually necessary or convenient. For instance, in ordinary arithmetic, a figure placed on the right of another means that it is to be multiplied by ten before adding it to the other. Thus 24 is $2\times 10+4$. We do not commonly use fractions in the same way: thus $2\frac{1}{2}$ 4 never stands for $2\frac{1}{2}\times 10+4$, or 29; but it might do so if we pleased; similarly $3\frac{1}{3}$ $2\frac{1}{2}$ might stand for $3\frac{1}{3}\times 10+2\frac{1}{2}$, or $34\frac{1}{2}$. We now propose the following

PROBLEM II. What is that number, consisting of two digits the sum of which is 10, and which is doubled by inverting the digits?

We see that 91 is not double of 19, nor 82 of 28, nor 73 of 37, nor 64 of 46. Therefore, with the restrictions on the decimal system usually adhered to in practice, the problem is impossible. What sort of answer then are we to expect if we reduce the problem to an equation? Let x and y be the digits in question: then

$$x+y=10, \quad \text{or} \quad y=10-x.$$

* For instance, place D on the right at a thousand times the distance of B from A. Then AD is to DB as 1001 to 1000, or in the proportion of $1+\frac{1}{1000}$ to 1, which is very nearly the proportion of 1 to 1.

The number formed by placing the digit x before y, is $10x+y$ (just as 24 is $2\times 10+4$, 58 is $5\times 10+8$, &c.); and when the digits are reversed, the number is $10y+x$ (just as 42 is $4\times 10+2$, 85 is $8\times 10+5$, &c.). By the problem, the second is double of the first; that is,

$$10y+x = 2(10x+y), \quad \text{or} \quad 20x+2y.$$

Therefore $10y-2y = 20x-x$, or $8y = 19x$.

But $y = 10-x$, or $8(10-x) = 19x$;

therefore, $x = \frac{80}{27} = 2\frac{26}{27}$, $y = 10-x = 7\frac{1}{27}$.

The answer therefore is, that if it be understood that none but the usual single digits shall be placed in the columns of units and tens, the problem is impossible; but that, if the method of writing numbers be extended, so that a fraction placed before a fraction shall be considered as meaning 10 times as much as when it stands alone, then the problem is possible, and the direct and inverted numbers are,

$$2\frac{26}{27}\,7\frac{1}{27} \quad \text{and} \quad 7\frac{1}{27}\,2\frac{26}{27}.$$

Here $2\frac{26}{27}\,7\frac{1}{27}$ means $2\frac{26}{27}\times 10+7\frac{1}{27}$ and is $36\frac{18}{27}$

$7\frac{1}{27}\,2\frac{26}{27}$ means $7\frac{1}{27}\times 10+2\frac{26}{27}$ and is $73\frac{9}{27}$

the second of which is double of the first.

We may then lay down the following: When a problem has a fractional answer, that answer can only be used on the supposition that any usual method of combining whole numbers, which is necessary in forming the equation, shall also be applied to fractions.

This principle, when introduced into common life, often induces us to suppose fractional parts of things which, in the strict and original meaning of the terms, have no fractional parts. Thus there is, properly speaking, no such thing as *half a horse*: there may be half the body of a horse (as to bulk or weight), half the *power* of a horse (which is but the half of a certain pressure), or there may be a horse of half the size, half the power, &c. of another horse—that is, we may halve any quality of a horse which can be represented by numbers; but not the complete idea which we attach to the word, because it contains notions which have no reference to number or quantity.

Nevertheless, we do not speak absurdly when we talk of a steam engine of the power of $20\frac{1}{2}$ horses, because it is there only the horses' power that we speak of, which can be numbered in pounds of weight; or of wolves eating half a horse, because we then speak only of a weight of flesh. Thus the problem—A horse can draw two tons; certain horses drew 5 tons, how many were there?—is absurd in the strictest sense, but not so if we confine ourselves to the only quality of a horse which is concerned in the problem, namely, power of drawing: and we may either say, there was $2\frac{1}{2}$ times as much power as a horse possesses, or the power of $2\frac{1}{2}$ horses. The same remarks would apply to the following: The reckoning came to £5, and the share of each person is £2, how many persons were there?—which cannot be solved in the strictest meaning of the words, but in which we may say that the whole reckoning is $2\frac{1}{2}$ as much as that of one person, or that of $2\frac{1}{2}$ persons.*

Problem III. There are two pieces of cloth of a and a' yards in length. The owner sells the same number of yards of both sorts at b and b' shillings per yard. If the remainders were then sold, of the first at c shillings a yard, of the second at c' shillings a yard, the total prices of the two pieces would be the same. What number of yards was first sold of each?

N.B. To avoid using too many letters, it is usual to employ the same letter with one or more accents,† to signify different numbers which have some common point of meaning. Thus a and a' are the lengths of two pieces of cloth, b and b' the prices per yard of the first pieces taken from each, and c and c' the prices per yard of the remainders. But a and a' differ as much in meaning as to the numbers they may stand for, as a and b; either may stand for any number named.

Let x be the number of yards first cut off from each piece; then the remainders are $a-x$ and $a'-x$ yards. The sums received for the first are therefore bx and $b'x$ shillings; for the remainders, $c(a-x)$ and $c'(a'-x)$ shillings. Consequently, by the conditions of the question,

* So when we say that the yearly mortality of a country is 1 out of $40\frac{1}{2}$ persons, we mean 2 out of 81.

† The symbol a' may be read *a accented*, a'', *a twice accented*, and so on. But *a dash*, *a two dash*, &c. are shorter, though not quite so correct in grammar.

$$\begin{aligned} bx+c(a-x) &= b'x+c'(a'-x) \\ bx+ac-cx &= b'x+a'c'-c'x \\ bx-cx+c'x-b'x &= a'c'-ac \\ (b+c'-b'-c)x &= a'c'-ac \\ x &= \frac{a'c'-ac}{b+c'-b'-c} \end{aligned}$$

Suppose we try to apply this to the following case: Let the number of yards be 60 and 80; let the prices of the number of yards taken from each at first be 10 and 9 shillings a yard; and let the prices of the remainders be 4 and 3 shillings a yard. We have then

$$a=60 \quad a'=80 \quad b=10 \quad b'=9 \quad c=4 \quad c'=3$$

$$x=\frac{a'c'-ac}{b+c'-b'-c}=\frac{80\times3-60\times4}{10+3-9-4}=\frac{0}{0}$$

an anomaly already discussed in page 25. We have there seen that it implies that any value of x will solve the equation, and this we shall find to be the case in the present instance. For if we return to the equation, we find it becomes

$$10x+4(60-x) = 9x+3(80-x)$$

or $$10x+240-4x = 9x+240-3x$$

or $$6x+240 = 6x+240$$

which is true for all values of x. Hence the answer is, that in this particular case the total prices of the two pieces are the same whatever number of yards be first cut off.

Let us now try another case. Let the pieces be 60 and 80 yards, as in the preceding; but let the first pieces cut off be sold at 5 and 4 shillings a yard, and the second at 2 and 3 shillings a yard; then

$$a=60 \quad a'=80 \quad b=5 \quad b'=4 \quad c=2 \quad c'=3$$

$$x=\frac{a'c'-ac}{b+c'-b'-c}=\frac{80\times3-60\times2}{5+3-4-2}=\frac{120}{2}=60$$

the number of yards cut from both is 60; that is, the *whole* of the first piece is taken, and 60 yards of the second, which are sold at 5 and 4 shillings a yard (giving 300 and 240 shillings). Then the remainder of the second (we need not mention the remainder of the first, which being nothing, brings nothing), 20 yards, sold at 3 shillings a yard, brings 60 shillings. The produce of the first is 300, of the second 240 + 60 shillings, both the same, as the problem requires.

The third case we will take is as follows: Let the pieces be 60 and 80 yards; let the pieces cut off be sold for 7 and 3 shillings a yard, and the remainders for 5 and 2 shillings a yard,

$$a = 60 \quad a' = 80 \quad b = 7 \quad b' = 3 \quad c = 5 \quad c' = 2$$

$$x = \frac{80 \times 2 - 60 \times 5}{7 + 2 - 3 - 5} = \frac{160 - 300}{1}$$

and contains an impossible subtraction. From the conclusions in page 18, we must suppose some wrong alternative has been employed in choosing x. But, on looking at the problem, no such thing appears; x yards are to be first sold of each. But the problems in page 18 were purposely* put in a form in which the alternatives were obvious; how then are we to widen the expressions used in stating this problem so that the problem, as we have given it, may be only one case out of two or more? Remember that we alter no number, but only the *quality* of the result. The seller begins with 60 and 80 yards of the two cloths in his possession, and ends with none of either, having in his pocket the same receipts from both pieces. Our problem says he first sells a certain number of yards from both, and the answer upon this supposition shews the problem to be impossible. We have been previously directed in such cases to alter the quality of the result; let us do this, and suppose he begins by buying the same quantity of both. We must preserve the condition that he begins with 60 yards of the first, and ends with none; therefore, if he begin by buying 10 yards more, he must sell the whole 70. If so, he also buys 10 yards more of the second sort, and sells the whole 90. But as we are to alter none of the numbers, but only change their names, if he buy more he buys at 7 (b) and 3 (b') shillings a yard; and when he gets rid of all he has (not all he has *left*, for that belongs to one particular alternative), he sells at 5 and 2 shillings a yard. Therefore the wider problem, of which the one proposed is one of the alternatives, is as follows:

* We may here remark that the extension of a problem appears natural or not according to the idiom of the language in which it is expressed. Thus, comparing together the case here given, and that in page 19, the former appears forced, because we have no very common word to denote either buying or selling as the case may be; the latter appears natural, because the words "give the date," implying asking for a time, either before or after a given epoch, are perfectly consistent with

EXTENDED PROBLEM.

There are two pieces of cloth a and a' yards in length. The owner *concludes a bargain respecting* the same number of yards of both sorts at b and b' shillings per yard. If *his stock of both* were then sold, of the first at c shillings a yard, of the second at c' shillings a yard, the *total results of the transactions in each sort of cloth* would be the same.

CASE FIRST PROPOSED.

There are two pieces of cloth a and a' yards in length. The owner *sells* the same number of yards of both sorts at b and b' shillings per yard. If *the remainders* were then sold, of the first at c shillings a yard, of the second at c' shillings a yard, the *total prices of the two pieces* would be the same.

The case first proposed leads, as before, to the equation

$$c(a-x)+bx = c'(a'-x)+b'x$$

or

$$x = \frac{a'c'-ac}{b+c'-b'-c}$$

but the general case leads to this equation only when the owner is the seller in the bargain mentioned. If he be the buyer, he first pays bx and $b'x$ for what he buys of each sort, and then sells the stocks $a+x$ and $a'+x$ at c and c' shillings per yard. Therefore $c(a+x)-bx$ is the balance in his favour from the first, and $c'(a'+x)-b'x$ from the second. The equation is

$$c(a+x)-bx = c'(a'+x)-b'x$$

or

$$x = \frac{ac-a'c'}{b+c'-b'-c}$$

---

the idiom of our language. But if we translated these two problems into a language, in which there was a word in common use, such as *trafficking*, to denote either buying or selling, and in which there was no usual way of asking for a date, without implying either before or after some other date, then the present extension would appear natural, and the one in page 19 forced.

The equations of algebra of course take no cognisance of such differences of idiom, which is generally considered one of the great advantages of the science, though some regard it as a defect. The student must decide this point for himself, when he has had sufficient experience of the advantages and disadvantages arising from such extensions.

which differs from the former, consistently with the rules in page 18, by an alteration of the sign of every term which contains x, and an inversion of the subtraction $a'c' - ac$ in the result. When

$$a = 60 \quad a' = 80 \quad b = 7 \quad b' = 3 \quad c = 5 \quad c' = 2$$

we have already tried the alternative of supposing the owner to be seller in the first transaction, and have found an impossible subtraction 160 — 300 in the result. If we now try the second alternative, and suppose him to be buyer, we shall get the rational answer 300 — 160 or 140; which will be found to satisfy the problem. We might enter upon various other cases of the same question, but we shall leave them for the present to the student, and state a problem which is in all respects analogous to the preceding, and presents similar alternatives in a different form. The equations of the two problems will be the same.

Problem IV. The paper is a map of a country, of which AB is a level frontier. All the roads rise from left to right, and fall from right to left, and all the miles mentioned are meant to be measured perpendicularly to the frontier on a level with it. CD is a parallel to the frontier (whether right or left of it is not stated), and T and V are towns on, above, or below (as the case may be), the line BD perpendicular to the frontier. P and Q are two frontier towns; R and S two towns, on, above, or below the points R and S, according to the position of CD. The roads rise or sink, as the case may be, a number of inches per mile for every level mile from the frontier, as follows:

From P to R b inches per mile.
From Q to S b' inches per mile.
From R to V c inches per mile.
From S to T c' inches per mile.

Now T and V are on the same level, BV is a level miles, and BT a' of the same. Required the distance BD and its direction.

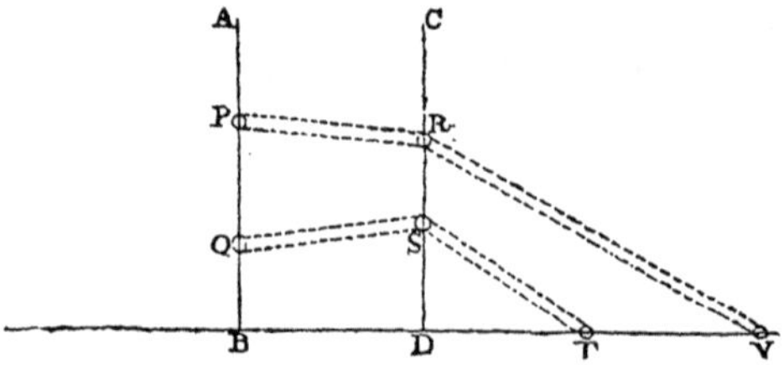

We have given this as an exercise, and shall merely point out the resulting equations on all the different suppositions. In all of them x stands for B D in miles.

1. If D lie to the left of B,

$$\left.\begin{array}{ll}\text{either} & c(a+x)-bx = c'(a'+x)-b'x \\ \text{or} & bx-c(a+x) = b'x-c'(a'+x)\end{array}\right\} \text{according as TV is above or below PQ.}$$

2. If D lie between B and T (as in the figure),

$$c(a-x)+bx = c'(a'-x)+b'x.$$

3. If D lie between T and V,

$$bx-c(x-a) = b'x+c'(a'-x) \text{ when } a' \text{ is greater than } a,$$

$$\text{or } bx+c(a-x) = b'x-c'(x-a') \text{ when } a \text{ is greater than } a'.$$

In the figure a is greater than a'.

4. When D lies to the right both of T and V,

$$bx-c(x-a) = b'x-c'(x-a')$$

$$\text{or} \quad c(x-a)-bx = c'(x-a')-b'x$$

the first when TV is above, the second when TV is below, the level of PQ. Only one of these seven equations can be altogether true; and as only the alternative in the set marked 3 is directly given in the conditions of the problem, six equations may need examination. But after the explanation of the anomalies (1), (2), and (3), any one of the six equations may be made to give the true answer. The anomaly (4) will be found to arise when $b+c'=b'+c$, and (5) when in addition to this, $a'c'=ac$.

The student may make this problem an illustration of every case which we have found to arise.

Before proceeding to other forms of equations, we shall now, having found impossible subtractions arise in the solutions of problems, and having seen the method of interpreting them, proceed to the investigation of rules, by which the interpretation may be deferred to any stage we please of the processes, and by which the symbols of impossibility may be used as if they were real numbers, without creating error.

CHAPTER II.

ON THE SYMBOLS OF ALGEBRA, AS DISTINGUISHED FROM THOSE OF ARITHMETIC.

SYMBOLS such as 50 — 70 (page 15), 200 — 500 (page 16), have received the name of *negative quantities*. This expression is not a correct one, because 50 — 70 is not a quantity, but the contradiction which arises out of writing down directions to do that which is impossible to be done. But we have seen, pages 14..19, that 50 — 70, when it is the answer to a problem, implies 20 things of some sort, as much as 70 — 50, but 20 things of a nature directly the contrary of the things first supposed; so that in forming the correct equations, these things diminish that which they increased in the incorrect one, and increase that which they diminished.

Conceive a problem consisting of various steps, such as would arise from joining two or more problems together, in such a way that the result of the first must be known before we can proceed with the second. If we solve the first problem, and find by an impossible result that we have chosen a wrong alternative, we should first retrace our steps, set the matter right, and when we have obtained an intelligible result, proceed to the second problem. But, we may ask, are there no rules by which this may be avoided, and by which we may continue the process, as if the result obtained had been rational? To try this question, we must *examine the consequences of proceeding with the symbols of impossible subtraction, to see what will come of applying those processes which have been demonstrated to be true of absolute numbers.*

If we look at a problem which presents alternatives, we shall see that not only will the answers be of different kinds, according as one or other alternative is the true one, but the methods in which the unknown answer (x) must be treated, in order to form the correct equation, are different in the two different cases. Thus, in page 15, Problem I., when x was years *after* 1830, the processes into which x entered were $50+x$ and $35+x$; but when x was years *before* 1830, these processes were $50-x$ and $35-x$. And this brings us to define what we mean by *different problems* as distinguished from *different alternatives of the same problem.*

This distinction has been already tacitly laid down; for, when we came to an impossible subtraction $a-b$, we always looked for that modification of the problem, which, not changing absolute numbers, but only the sense in which they were taken, would have given $b-a$ as the answer. Or when we came to Anomaly I., page 12, in which not the answer, but the verification of it, was impossible, we chose that modification which, without altering absolute numbers, inverted the impossible subtractions in the equation. Hence, without perceiving it, we have been led to make use of the following definition, [in which, however, it must be recollected, that we have obtained it entirely from equations of the first degree, and that it must therefore be considered as limited to such equations, and possibly capable of extension for equations of higher orders.] It is convenient to consider those problems only as different alternatives of the same problem, in which, 1. the absolute numbers employed are the same; 2. the equations only differ in having inverted terms with different signs (such as $-\frac{100-x}{2}$ instead of $+\frac{x-100}{2}$ in page 13), or else in having the signs of the unknown quantity altered throughout (as $50-x=2(35-x)$ instead of $50+x=2(35+x)$ in page 15); 3. the answers are either the same in both, or else only differ in the inversion of a subtraction (such as 70—50 instead of 50—70, in p. 15).

We might propose a problem appearing to have alternatives, but which, in this view, is two different problems combined together. For instance, A has £60, and is to receive the absolute balance that appears in B's books, whether for or against B; but C, who has £200, is to take B's property, and pay his debts. After doing this it is found that C's property is three times that of A. What is the absolute balance for or against B?

If B have x pounds, the equation is

$$3(60+x)=200+x \quad \text{or} \quad x=10$$

If B owe x pounds, the equation is

$$3(60+x)=200-x \quad \text{or} \quad x=5$$

We have here two different equations not reducible one to the other by any of the changes allowed in the preceding definition; and therefore, so far as the preceding is considered as reducible to equations of the first degree, the two cases are distinct problems.

The step we now make is to apply the rules of algebra to the symbols of impossible subtractions, and afterwards to proceed correctly with the inverted subtractions, that we may see, by comparing the results, whether the errors committed may be corrected or not by simple and general rules. And since we know that such a subtraction as $3-7$ will, when set right, be $7-3$ or 4, let us denote it for the present by $\bar{4}$, in which the bar written above 4 is not* a sign of subtraction, but a warning that we are using the inverted form $3-7$ instead of $7-3$. Thus we might say, that $10-14$ should also be denoted by $\bar{4}$; but here we must stop until we have some further assurance upon this point; for we cannot as yet *reason* upon such symbols as $3-7$ and $10-14$, since they represent no quantity imaginable, and we have not yet deduced rules. All we can do is to go to the source from whence they came, and see whether, by the same means which gave $3-7$, we might have got $10-14$ in its place.

Suppose a problem, wrongly expressed or understood, gives $2x-x=50-70$ (as in page 15). If we take the correct equation $50-x=2(35-x)$, we may add any quantity to both sides. Say we add a to both sides, which gives

$$(50+a)-x = (70+a)-2x$$

the solution of which is

$$2x-x = (70+a)-(50+a) = 20$$

If we had taken the incorrect form $50+x=2(35+x)$, and added a to both sides, we should, by what we suppose to be strict reasoning, till the result undeceives us, arrive at the conclusion $x=(50+a)-(70+a)$. And this is what we want to ascertain, namely, that just as in the rational form of an equation, we may by previous legitimate alterations obtain the answer in the form $70-50$, $71-51$, $72-52$, &c.: so, in the irrational form, we may, by the same sort of process, obtain $51-71$, $52-72$, &c., instead of $50-70$. And we will not say, that $51-71$ is *equal* to $50-70$, because our

* The sign of subtraction is a bar written *before* the quantity to be subtracted. The present is not a sign of subtraction any more than the bar between the numerator and denominator of a fraction.

notion of equal (as far as we have yet gone) applies only to magnitude, number, bulk, &c. &c.; but because any equation which gives 50 — 70, might also have been made to give 51 — 71, &c., we will call these equivalent to 50 — 70, meaning by the word *equivalent*, that the first may stand in the place of the second, or be substituted for it, without producing any error when we come to correct the result, or any set of operations for which correct substitutes cannot be found. Thus, then, we say, that 0 — 1, 1 — 2, 2 — 3, &c., are all equivalent, and may be represented by $\bar{1}$; $a - (a+b)$ and $(a+z) - (a+c+z)$ are equivalent, and are represented by $\bar{c}$; the rule always being;—invert the subtraction, and place a bar over the result.

Thus, if we obtain such an equation as

$$x + a + b = 0$$

which is the form most obviously impossible of all, we shall, *by rules only*, obtain the expression

$$x = 0 - (a + b)$$

which we signify by $\overline{(a+b)}$

In considering equations of the first degree, we may confine ourselves to the rational form $x - a = 0$, and the irrational form $x + a = 0$. For to these all others can be reduced. For instance (page 5) equation 4 is reduced to

$$x - \frac{12}{13} = 0$$

and the first equation of Prob. I., in page 15, is reduced to

$$x + 20 = 0$$

We now find equivalent forms for

$$\bar{a} + \bar{b} \qquad \bar{a} - \bar{b} \qquad \bar{a} \times \bar{b} \qquad \frac{\bar{a}}{\bar{b}}$$

The first will arise from such an equation as the following:

$$x + (p + a) - p + (q + b) = q$$

from which, if we solve it before observing that it is impossible, we have

$$x = p - (p + a) + q - (q + b) = \bar{a} + \bar{b}$$

But the same rules of solution will also give

$$x = p + q - (p + q + a + b) = \overline{(a + b)}$$

In which the bar over $a+b$ signifies that we have attempted to subtract from a quantity another which is greater than itself by $a+b$. Or we have

$$\bar{a}+\bar{b} \text{ is equivalent to } (\overline{a+b})$$

Similarly the equation

$$x+(p+a)-p+q = (q+b)$$

gives $$x = p-(p+a)-\big(q-(q+b)\big) = \bar{a}-\bar{b}$$

following rules only. But this equation is not always impossible, for it is equivalent to

$$x+p+a-p+q = q+b$$

or $$x+a = b \text{ which gives } x = b-a$$

Therefore, $\bar{a}-\bar{b}$ is correctly $b-a$ when b is greater than a; when b is less than a, it is $(\overline{a-b})$. We may also give the equivalent form $b+\bar{a}$, which follows from the attempt to solve correctly.

$$x+(p+a)-p = b$$

in the form $$x = b+\big(p-(p+a)\big)$$

We have as yet obtained no such expressions as

$$\bar{a}\bar{b} \text{ or } \frac{\bar{a}}{\bar{b}}$$

but we shall now shew that these arise from inattention to the equation, not to the problem. That is, we shall deduce $\bar{a}\bar{b}$ and ab, $\frac{\bar{a}}{\bar{b}}$ and $\frac{a}{b}$, or similar forms, from the same equation, whether that equation be true or false.

If p be greater than q, and c greater than d, the multiplication of the rational expressions $p-q$ and $c-d$, and the attempt at multiplication of the irrational expressions $q-p$ and $d-c$, give the same result, as follows:

$$\begin{array}{lll} & p-q & q-p \\ & c-d & d-c \\ \hline & pc-qc & qd-pd \\ & \quad pd-qd & \quad qc-pc \\ \hline \text{Subtract} & pc-qc-pd+qd & qd-pd-qc+pc \end{array}$$

which are the same in every thing but the order of their terms: Consequently, the equation

$$x + qc + pd = pc + qd$$

might, by inattention, be thus solved

$$x = pc + qd - qc - pd = (q - p)(d - c)$$

where the latter should have been

$$x = (p - q)(c - d)$$

whence, if $p - q$ be a and $c - d$ be b, the expression $\bar{a}\bar{b}$ may be obtained instead of the quantity ab.

We have already seen, in page 19, how it may happen that $\frac{\bar{a}}{\bar{b}}$ is introduced in the place of $\frac{a}{b}$. Let p be greater than q, and c greater than d; then

$$cx + q = dx + p$$

correctly solved, gives

$$(c - d)x = p - q \qquad x = \frac{p-q}{c-d}$$

incorrectly solved, gives

$$(d - c)x = q - p \qquad x = \frac{q-p}{d-c}$$

whence, if $p - q$ be a, and $d - c$ be b, $\frac{\bar{a}}{\bar{b}}$ is equivalent to $\frac{a}{b}$

We have thus determined either equivalent forms or corrections, in all the cases which arise from the equation, considered without reference to the problem from which it was derived. We shall now consider both, taking as a basis what we have ascertained by examples, namely, that the misconception which produces $\bar{a}$ instead of a, also causes us to add where we should subtract, and subtract where we should add, in forming our equations.

Suppose we have obtained $\bar{a}$ as the result of an equation, and that the next step (if all before had been correct) would have been to add $\bar{a}$ to c, or to find $c + \bar{a}$.

Correction and Process. The true result is a, but the same misconception by which $\bar{a}$ was produced, has made us suppose this quantity was to be added where it was in reality to be subtracted, and *vice versâ* (pages 15–18); therefore the next step corrected is $c - a$ or $c + \bar{a}$ is equivalent to $c - a$.

Trial of the incorrect Process. $\bar{a}$ is the representative of $z-(z+a)$, and $c+\bar{a}$ (applying rules only) is $c+z-(z+a)$, or $c+z-z-a$, or $c-a$, which has been shewn to be correct.

Similarly, where $c-\bar{a}$ occurs, we know that the true answer is a, but that the misconception which produced $\bar{a}$ leads us to suppose it should be subtracted where it should be added; therefore $c+a$ is the true result. The incorrect process gives $c-\big(z-(z+a)\big)$, or $c-z+(z+a)$, or $c-z+z+a$, or $c+a$. We have, then, these two rules:

$$c+\bar{a} \text{ is equivalent to } c-a$$
$$c-\bar{a} \quad \cdots\cdots\cdots\cdots \quad c+a$$

By similar reasoning $(p-\bar{a})\times(q-\bar{b})$ is equivalent to $(p+a)(q+b)$. The incorrect process gives $pq-q\bar{a}-p\bar{b}+\bar{a}\bar{b}$. If we consider $\bar{a}$ and $\bar{b}$ as resulting from the misunderstanding of a problem, we must take some such problem as a guide. Let it be the following:

A and B have respectively 4 and 5 pounds. One of them loses a bet to the other, after which the product of the number of pounds belonging to each is 18. Which lost, and how much?

If we suppose A lost x pounds, the equation evidently is

$$(4-x)(5+x)=18$$

But if we suppose A gained x pounds, the equation will be

$$(4+x)(5-x)=18$$

First Alternative.	Second Alternative.
$4-x$	$4+x$
$5+x$	$5-x$
$20-5x$	$20+5x$
$4x-xx$	$4x+xx$
Add $20-x-xx$	Subtract $20+x-xx$
$20-x-xx=18$	$20+x-xx=18$
$xx+x=2$	$xx-x=2$

From this we see, that in this instance (and trial proves the same in others of the same kind) the correct equations belonging to different alternatives have different signs in the terms which are multiplied by x only once (or contain the first power of x), or (page 1) are of the first degree with respect to x; but that the terms which contain xx, or are of the second degree with respect to x, have not different signs. And the same may be proved of terms containing the product of two unknown quantities x and y, which are not the same; for instance, in the equations

$$(4-x)(5+y) = 18$$

and

$$(4+x)(5-y) = 18$$

If we take a term such as $\bar{a}\bar{b}\bar{c}$, we shall find that, on correction, the sign preceding it must be changed; that in such a term as $\bar{a}\,\bar{b}\,\bar{c}\,\bar{d}$, the preceding sign is not changed: the rule being, *change* the sign where there is an *odd* number of factors to be corrected. It is indifferent whether the factors to be corrected be in the numerator or denominator: if, for example, we take $\frac{\bar{a}\,\bar{b}}{\bar{c}}$, we may, by deducing $\frac{1}{x}$ instead of x from the equation which gave $\bar{c}$, find $\frac{\bar{1}}{c}$ instead of $\frac{1}{\bar{c}}$, as follows. Suppose the equation which gives $\bar{c}$ to be

$$2x+(c+z) = x+z:$$

this, following only rules, may be finally represented either by

$$2x-x = z-(c+z) \quad \text{or} \quad x = \bar{c}$$

or by

$$x-2x \quad \text{or} \quad (1-2)\,x = c+z-z = c$$

that is,

$$(\bar{1})x = c \qquad (\div)\,cx \qquad \frac{\bar{1}}{c} = \frac{1}{x}$$

Tracing the consequences of this result, we find that the uncorrected form $\frac{\bar{1}}{c}$ is equivalent to $\frac{1}{\bar{c}}$. Hence the term $\frac{\bar{a}\,\bar{b}}{\bar{c}}$ is equivalent to $\bar{a}\,\bar{b}\times\frac{\bar{1}}{c}$, or $\frac{\bar{1}\,\bar{a}\,\bar{b}}{c}$, which contains all the uncorrected factors in the numerator.

If we now resume $(p-\bar{a})\,(q-\bar{b})$, of which the continuation of the incorrect process is $pq-p\,\bar{b}-q\,\bar{a}+\bar{a}\,\bar{b}$, we must conclude, that, in the terms of the first degree with respect to $\bar{a}$ and $\bar{b}$, the signs must be altered; but that in the term $\bar{a}\,\bar{b}$ the sign preceding it must not be altered; so that the correction gives

$$pq+pb+qa+ab$$

But this is the same as would have arisen if the process had been corrected one step earlier, as in page 50; that is, if $(p+a)(q+b)$ had been written at once for $(p-\overline{a})(q-\overline{b})$.

It is not necessary to go through all the individual cases that might arise. We have found in all that the common rules of algebra may be applied without error to the expressions of impossible subtractions; that is to say, the correction may be deferred as long as we please without introducing error, provided that, when at last the correction is made, the following rule be observed:—In correcting any term, change the symbols of impossible subtraction by substituting the absolute number resulting from the real subtraction [thus, put 3 for $\overline{3}$, or $(z+3)-z$ for $z-(z+3)$]; change the sign preceding all such terms as require an odd number of such corrections; keep the sign of such as require an even number. Repeat this as often as may be necessary. If the final result be then rational, the problem has been rightly understood, and the mistakes have arisen from inattention to the processes which come between the statement and the result; but if the result after correction be still an impossible subtraction, then the problem has been misunderstood if there be alternatives in it, or is itself a wrong alternative of some more general statement.

As an example, let the final result of a set of operations be

$$\frac{abc+dd}{ac-d}+\frac{ab}{cd}-ce$$

Let it be then discovered that a, b, and e, are not rational expressions, and let them be found to be made, a by attempting to subtract a number from 1 less than itself, b by attempting to subtract a number from 3 less than itself, and e by attempting to subtract a number from 5 less than itself. That is, let a, b, and e, be represented by $\overline{1}$, $\overline{3}$, and $\overline{5}$. Let c and d, which are rational, be 2 and 6. Then, the preceding expression, before correction, is

$$\frac{\overline{1}\times\overline{3}\times 6+2\times 2}{\overline{1}\times 6-2}+\frac{\overline{1}\times\overline{3}}{6\times 2}-6\times\overline{5}$$

the first stage in the correction of which is

$$\frac{18+4}{-6-2}+\frac{3}{12}+30$$

but, $-6-2$ (page 47) is signified by $\bar{8}$; and the necessary correction, according to the preceding rule, gives

$$\frac{3}{12}+30-\frac{22}{8} \text{ or } 27\tfrac{1}{2}$$

which is the result which would have been obtained, had each correction been made in its proper place.

But the preceding form $\bar{a}$ is not made use of by algebraical writers, and is introduced here not to remain permanently, but to avoid using the sign of subtraction, to appearance at least, in two different senses. If we follow rules, without observing where they lead us, we should obtain processes of the following sort:

$$3-8=3-(3+5)=3-3-5=0-5$$

and, as in $0+5$, it is not necessary to the meaning to retain 0, we might, by imitation, write -5. This is what we have hitherto written $\bar{5}$. And we shall find, in all the intelligible properties of the sign $-$, a close similarity between -5 (properly placed in an expression) and the legitimate rules by which it is treated, and $\bar{5}$, an uncorrected misconception, with the rules for obtaining a correct result.

In fact, if we apply rules to $+$ and $-$ as they are applied when the quantities concerned are rational, we find no distinction between $\bar{a}$ and $-a$, though we have made one until we could establish the points of similarity. For instance, $b-\bar{a}$ when corrected is $b+a$. And $b-(z-a)$, when z is greater than a, gives $b-z+a$; so that a, rationally used with two negative signs before it, gives $+a$ in the result. Apply this rule to $b-(-a)$ and it gives $b+a$; so that $b-\bar{a}$ is corrected by the application of no other rule than considering $\bar{a}$ as $-a$, and applying the rules which, in the Introduction, are shewn to apply in *possible* subtractions. The same will be found in every other case yet stated.

In further illustration, let $a=b-c$, let $c=d-x$, let $x=y-v$, and let $v=t-z$. We have, then,

$$\begin{aligned} a=b-(d-x) &= b-\big(d-(y-v)\big) \\ &= b-\Big\{d-\big(y-(t-z)\big)\Big\} \quad \ldots\ldots\ldots \text{(A)} \\ &= b-\Big\{d-(y-t+z)\Big\} \\ &= b-\Big\{d-y+t-z\Big\}=b-d+y-t+z \end{aligned}$$

On looking at this result, we see that z is preceded by $+$; in the expression (A) it is four times under the negative sign. t is preceded by $-$, and is under three negative signs. y is preceded by $+$, having been under two negative signs. Consequently, those terms are negative which are under an odd number of negative signs; and positive, which are under an even number.

Now, let us suppose that we have obtained from an equation an incorrect value $0-a$, denoted by $\bar{a}$, as before. Let the result of some succeeding process lead to $0-\bar{a}$, which we will denote by $\bar{\bar{a}}$. A third process leads to $0-\bar{\bar{a}}$, which we denote by $\bar{\bar{\bar{a}}}$, and so on. At every step, therefore, a new misconception has been introduced: but, as we shall proceed to shew, the repetition of the error any even number of times does not shew itself in the result. The first irrational answer must arise from the attempt to determine x from an equation, which reduces itself to $x+a=0$. Let a new process be then entered upon to determine y, and let it lead to $y+x=0$. The equations which we should have got by proceeding correctly are $x-a=0$ and $y-x=0$, which lead to $y=a$. But this is the same as we should get from the incorrect equations $x+a=0$, $y+x=0$, by rules only. For, subtracting the first from the second, we have $y-a=0$, or $y=a$. Again, proceeding to find z, suppose we get $z+y=0$. From the correct equations in the first column

$$\begin{array}{ll} x-a=0 & x+a=0 \\ y-x=0 & y+x=0 \\ z-y=0 & z+y=0 \end{array}$$

we get $z=a$; not so now from the incorrect equations (observe that their number is odd). For, from rules only, by adding the first to the third we get $z+x+y+a=0$, from which, subtracting the second, we find $z+a=0$. And thus we might proceed with more equations. Now, observe that the first equation gives $x=0-a$ or $\bar{a}$, the second $y=0-x$ or $0-\bar{a}$ or $\bar{\bar{a}}$, the third $z=0-y$ or $0-\bar{\bar{a}}$ or $\bar{\bar{\bar{a}}}$, and so on. Hence, from the preceding, any even number of errors of this kind corrects itself, any odd number requires correction.

We now return to the expression

$$b-\left\{d-\left(y-(t-z)\right)\right\}=b-d+y-t+z$$

which is rational when t is greater than z, y greater than $t-z$, &c.

We have already applied mere *operations* to irrational cases to see what would come of them; we now apply direct *reasoning* knowingly to an irrational case, with the same intention. Not that any new step is now made; for in applying operations, we always tacitly employ the reasoning by which the rule of operation was demonstrated. It is only allowable to write an instance of $a-(b-c)=a-b+c$ in cases where we can return to the demonstration (see the Introduction), and make that demonstration apply to the particular case in hand. We place the example of too general reasoning in brackets.

[Since the preceding equation is generally true, it is true when $t=0$, $y=0$, $d=0$, and $b=0$. But the preceding equation, omitting the term 0 wherever it occurs, because 0 neither increases nor diminishes the value of an expression, becomes

$$-\left\{-\left(-(-z)\right)\right\}=z$$

therefore z is not altered by being preceded by *four* negative signs, and the same may be proved of any *even* number]

But the preceding reasoning is obviously incorrect, if seriously meant as a proof, and not as an experiment; for the equation $a-(b-c)=a-b+c$ is tacitly applied to the case where $b=0$. We will go over the general proof, which ought to apply to the particular case, putting the two together.

GENERAL PROOF.	PARTICULAR APPLICATION.
(True when b is greater than c.)	(Not admissible as reasoning in any case.)
$b-c$ is to be taken from a. If from a we take away b, giving $a-b$, we have too little, because only a part of b should have been taken away, namely, as much as is left after it has been lessened by c. Therefore, c too much has been taken away, and the true result is $a-b+c$.	$0-c$ is to be taken from a. If from a we take away 0, giving $a-0$, we have too little, because only a part of 0 should have been taken away, namely, as much as is left after it has been lessened by c. Therefore, c too much has been taken away, and the result is $a-0+c$ (that is, $a+c$).

The application of the reasoning needs no comment. *Nothing* has been taken away from a quantity, and yet it has been too much diminished—only a part of nothing should have been taken away, and *nothing* should have been previously *lessened.* We will now subjoin the version we give of the equation $a-(-c)=a+c$, derived from the preceding part of this chapter.

We have examined particular cases, and have always found that $0-c$ is the result of a mistake of one particular sort, namely, that the quantity represented by c is of the opposite nature to what we supposed it to be. And we have also found by examination, that as to addition and subtraction, all the operations have been inverted; we have added where we should have subtracted, and subtracted where we should have added; to which additional misapprehension we should have remained subject, had we not come to the irrational expression $0-c$. Therefore, instead of $a-(0-c)$, we remember that $0-c$ should be $c-0$ or c, and in $a-(0-c)$ the first $-$ is also the result of misconception, and should be $+$. Therefore $a-(-c)$ is correctly written (not *is equal to*) $a+c$. Thirdly, we have found by examination, that if we continue any series of processes with the effects of the misconception uncorrected, we shall not introduce any errors but those which may be corrected, at any period we please, by the use of one simple rule, namely, apply no other rules except those which have been deduced in the case of rational expressions.

It was found out that the rules of algebra might be applied without error to symbols of impossible subtraction, before the cause* of so singular a circumstance was satisfactorily explained. The consequence was, that many such reasonings as those in page 55 were universally received, and a language adopted in consequence which, as long as words have their usual meaning, is absurd.

But, at the same time, every one must see that words are them-

* I am far from asserting that the view I have taken will be easy, or that it is the only one which might have been given as satisfactory to those who can understand it. But I think that the matter of it, independent of the method of stating it, must be considered at least of incontrovertible logical soundness. I am aware that many will think the connexion of $-(-a)$ and $+a$ to be more *necessary* than I have attempted to shew it to be; and in the higher view of the subject, which no beginner could understand, it may be so; but I think the exclusion of false analogies (to which the student is very subject in this part of the science) of more importance than the establishment of true ones. It will be easier for the pupil hereafter to acquire new ideas of relation, than to get rid of any he now acquires.

selves symbols of our own making, over which there is unbounded control consistently with reason, provided only that what we mean by every word be distinctly known, so that we shall not draw conclusions from one meaning of a word, and then apply those conclusions to other meanings.*

We have made some additions to common arithmetic, and have found uses for symbols which were never contemplated in that science. It is sufficient that we have demonstrated the uses of these symbols; it now remains to find words by which to express the operations we shall employ.

There are two ways of proceeding.

1. Whenever we want to signify an operation which is not wholly arithmetical, we may invent a new term. This would load the science with difficult words, which, after all, would only have the effect of banishing the arithmetical words, and substituting others in their places; for we cannot know whether we are proceeding with or without the misconceptions explained in this chapter, until the end of the process. We should therefore be obliged always to use the newly invented terms, to the exclusion of the others.

2. We may alter the meaning of the words already in use by *extending* them; that is, allowing them to mean what they already stand for, *and more.* In common language we are already well used to something like this, and, whenever we want a word to express one object, are in the habit of using one which belongs to another object having some resemblance to, or connexion with, the one we are to name. By this means, there is a word in the English language which stands for a receptacle, a seat, a small room, a small house, a plant, and a blow. But this is forming names by resemblance only, not by extension. We might instance the latter in the words *arm, mark, plain,* &c.; but it is to the sciences only that we must look for examples

* As an illustration of our meaning: The word *square,* in algebra (as we shall see when we come to it), is made to mean *a number multiplied* by itself; in geometry, it means a well-known species of four-cornered figure. How this word came to have two meanings so different, the student will see if ever he studies the history of algebra, and will guess when he comes to apply algebra to geometry. But in the mean time, nothing that is proved of the square in algebra is to be therefore taken for granted of the square in geometry. The same word with two different meanings is the same as two different words.

of intentional extension, correctly managed. We take one out of the numerous instances which natural history affords. It is found convenient to divide all animals into classes, comprising in each class those which have certain common arrangements of teeth or other members. One particular class of animals contains the cat, which is the best known of its class. Instead of inventing a new word to signify this class, containing the cat, lion, tiger, panther, &c., they are all called *cats*, and each has a particular additional name to distinguish it. Thus, the common cat is *felis catus*, the lion is *felis leo*, the tiger is *felis tigris*, the panther is *felis pardus*, &c. Observe that what is here done is, to make the word *felis* (Latin for *cat*) mean less than in common discourse, and imply not the common animal, or any animal which agrees in all respects with it, but any animal which agrees with it only in those arrangements which are considered the distinguishing marks of the class. Consequently, by limiting the ideas which the word is meant to imply, the number of objects which come under it is extended. And no mistake could arise by this means when one zoologist speaks or writes to another; though a third person, not acquainted with their meaning, might think they believed that a cat could run off with and devour a man.

Similarly, in algebra, we have terms which are well understood in arithmetic, and processes which we find carry us beyond the object of arithmetic, which is absolute number. But, both in the processes of algebra, and in those of arithmetic, there are resemblances which will make it convenient to classify together those which follow the same rules; and, in giving the names to classes, we shall, as previously described, limit the definition of the arithmetical terms so as to name the whole class by the arithmetical name. And when we speak of the process of arithmetic to which we have been accustomed, we shall prefix the word *arithmetical*. Thus, by *arithmetical* addition, we mean the simple increase of one absolute number by another. But this will be, as we shall see, only one case of *algebraical* addition, or, as we shall call it, addition. *And, once for all, observe that in future every term has its extended or general algebraic meaning, except when the word* arithmetical *is prefixed.*

We shall now proceed to the limitations of the notions contained in the terms, or the extensions of the cases which come under them, whichever it may be called.

1. *Quantity* is applied in *arithmetic* to any number or fraction.

In *algebra*, it is any symbol* which results from the rules of calculation. We have the first effects of this in the following proposition.

Quantities (as far as we have yet gone) *are either positive or negative.* Arithmetical *quantities are all positive.*

The latter part requires this remark, that $+6$ is more than the symbol of 6; it is the symbol of one particular way of obtaining 6, namely, $0+6$, or adding 6 where there was nothing. But $3+3$ is not identical with $+6$ in the *operations indicated*, but only in the *result obtained;* just as $(a+x)(a-x)$ and $aa-xx$, which are not the same operations, always give the same results.

Positive and negative quantities are of diametrically opposite significations; if $+a$ *be a gain of* £a, $-a$ *is a loss of the same; if* $+a$ *be a loss of* £a, $-a$ *is a gain of the same; if* $+a$ *be length measured northward,* $-a$ *is the same length measured southward; and so on.*

This distinction has been sufficiently dwelt upon in a different form. If we suppose a gain, and the answer to our supposition (derived from a rational equation) is that the gain is a, then we were right in our supposition; but if the answer be $-a$, then we were wrong in the supposition, and we ought to have supposed a loss of £a. The extension here made consists in making the symbol of an error stand for the correction of that error. On this we must remark, that the extension avoids confusion only because there is but one way of making the wrong supposition right. If $-a$ might indicate mistakes of more than one kind, it would be wrong to let $-a$ stand for the correction of one species, when the problem might oblige us to choose another.

The application of algebra to geometry immediately suggests this sort of extension. Let A be a point, in a straight line indefinitely extended both ways; measure AB to the right of A, and from B

* A symbol is any thing which can be placed before the mind as a representation of any other thing. The term *quantity* is not applied according to its original meaning, even in arithmetic. This 2 is not a *quantity* (unless it be of printer's ink), but a *symbol*, which denotes that a certain quantity, represented by 1, has been taken twice. Thus, $3-5$ is a rational symbol, but not a symbol of arithmetical quantity, because it proposes operations which contradict the arithmetical meaning of 3, $-$, and 5.

measure BC to the left of B. If AB be greater than BC, the proper description of the position of C is "at a distance AB — BC on the right of A." But if BC be greater than AB, and we still say the same of C, we are warned of some mistake by the impossible subtraction, and we see immediately that the proper designation of C's position is "at a distance BC — AB on the left of C."

In conversation and writing, figurative extensions are so common that they have sometimes been appealed to as a justification of the processes we have been investigating. In strict propriety, there is a repetition of ideas in the phrase "to gain a gain," and contradiction in "to lose a gain," in which the word "to lose" is used in the common sense of "not to get." But the most obvious analogy is in the words "to gain a loss," which is an ironical term applied to one who loses where he thought he should gain. And when we say, "darkness went away," instead of "light came," we make a mistake* in a matter of fact which bears a close analogy to that which we have considered.

In the application of mathematics to physics, we are liable to the error of imagining that a phenomenon may arise from matter being added to other matter, when in fact it arises from matter being taken away, and *vice versâ*. This has happened in several instances, of which we will cite two of the most remarkable.

1. If glass be rubbed against leather in dry weather, both substances acquire power to attract small pieces of matter, which power is called *electrical* attraction, or *electricity*. It was at first supposed that friction made the leather communicate some fluid to the glass, so that the glass had more than its natural share, and the leather less.

* These phrases are not introduced as illustrations or confirmations, but precisely the reverse. They are an impediment, because the student may by them be led to imagine that he sees reason in the use of the negative sign, independently of the proof given that it is merely a convenient method in correction of unavoidable misconceptions. To warn him that he has not (from this work at least) any evidence of the propriety of negative quantities, other than that which he gets from observing what will come of using them, is the object of this note. If a meaning is to be given to a term, which in its original use it will not bear, the more repugnant the phrases employed are to common ideas, the better in one respect; because the less the student can find any thing like them in his mother tongue, the more likely will he be to fasten upon them the explanation which they are meant to bear, and no other.

Consequently, the glass was said to be *positively*, and the leather *negatively*, electrified; and the phenomena of the latter were supposed to be caused by the subtraction of something from it. But succeeding experiments shewed it to be much more likely that the friction of the two substances separated a compound fluid, substance, agent, or whatever it may be called, into two distinct component parts, having this quality, that when united in their natural proportions they attract nothing, but that either, when separated from the other, shews it by attraction. These were called the *vitreous* and *resinous* electricities, because friction gave the first to *glass*, and the second to *resin* (as was found). But many still retained the old names of *positive* and *negative* electricity; and this produced no inconvenience, because what we may call the mathematical phenomena of electricity remained the same on both theories, it being exactly the same in calculating effects, whether we suppose the cause of the effect to be removed, or a sufficient quantity of something which destroys the effect to be added.

2. In burning a candle in a close vessel of air, it was observed that the air soon became incapable of allowing the process of burning to continue, and that the air produced was not fit to breathe. That an alteration had taken place was then certain; and it was supposed that the burning candle gave out a fluid which mixed with the air. This fluid was called *phlogiston* (thing which makes flame). Therefore the effect of burning on air was supposed to be the *addition of phlogiston*. But it was afterwards discovered that in fact something is taken from the air when a body burns, which something is oxygen, found by other means to be a part of the mixture called air. Hence, the effect of burning is the *subtraction of oxygen*. And if any chemical calculation made on the theory of phlogiston were required to be set right, it might be done on the supposition that $+a$ of phlogiston is $-a$ of oxygen, with the rules laid down in this chapter.

2. *Addition* and *Subtraction*. The first term means the forming two expressions into one, retaining the proper signs; the second, forming two expressions into one by altering the sign of the one which is said to be subtracted. The following examples will shew that arithmetical addition and subtraction are particular cases of the term.

$$3+(5-2) \quad \text{or} \quad 3+(0+5-2)=3+5-2$$
$$8-(5-2) \quad \text{or} \quad 8-(0+5-2)=8-5+2$$

But addition and subtraction include such as the following:

$$-3+(-5) \text{ is } -3-5 \text{ or } -8$$

$$-3-(-5) \text{ is } -3+5 \text{ or } +2$$

3. *Equal.* Any two algebraical expressions, of which the one may be substituted for the other without error, are called equal, and $=$ is written between them. This, as before, includes arithmetical equality; for $5+3$, which is arithmetically equal to 8, may be substituted for 8. But the algebraical term also applies to $3-7$ and $10-14$, page 46, to $a+(-b)$ and $a-b$, and so on; and the term will afterwards apply to still wider cases. For instance, we shall come to a species of misconception, which will give

$$1-1+1-1+1-1+ \text{ \&c. continued for ever.}$$

where the true result is $\frac{1}{2}$. This will be thus represented:

$$\tfrac{1}{2} = 1-1+1-1+ \text{ \&c. } \textit{ad infinitum}.$$

4. *Greater* and *less; increase* and *decrease.* The extension of the words addition and subtraction requires also the extension of these. The symbols of arithmetic are,

$$0 \quad 1 \quad 2 \quad 3 \quad 4 \quad 5, \text{ \&c.}$$

and intermediate fractions; and the greater of any two is that which comes on the right. The numerical symbols of algebra are,

					ARITHMETICAL.					
....	-4	-3	-2	-1	0	$+1$	$+2$	$+3$	$+4$	&c.

in which the addition proceeds throughout as in the arithmetical series by the (here algebraical) addition of $+1$. For

$$-4+1 = -3 \qquad\qquad -1+1 = 0$$

$$-3+1 = -2 \qquad\qquad 0+1 = +1$$

$$-2+1 = -1 \qquad\qquad +1+1 = +2$$

Let the definition of *greater* and *less* remain the same on both sides of the line; namely, that of any two quantities, the one which falls on the right is the greater. Thus -1 is called greater than -2, $+2$ is greater than -1, and so on.

Hence, with the extended meaning of the words, we have the following proposition:*

* This is the proposition which has startled so many beginners, and not without reason, considering that they have frequently been introduced to it without any warning that *greater* and *less* have not their

All positive quantities are greater than nothing; all negative quantities are less than nothing. Of two positive quantities, that is the greater which is arithmetically the greater; of two negative quantities, that is the greater which is arithmetically the less.

The extended terms *increase* and *decrease* will follow *greater* and *less*. Quantity is increased when it is made greater, and decreased when it is made less. But the word *smaller* is always allowed to retain its arithmetical meaning, without extension.

N.B. We have now separated *increase* from *addition*, &c.

Addition of .. $\left\{\begin{matrix}\text{positive}\\ \text{negative}\end{matrix}\right\}$ quantity causes $\left\{\begin{matrix}\text{increase}\\ \text{decrease}\end{matrix}\right\}$

Subtraction of $\left\{\begin{matrix}\text{positive}\\ \text{negative}\end{matrix}\right\}$ quantity causes $\left\{\begin{matrix}\text{decrease}\\ \text{increase}\end{matrix}\right\}$

The following propositions are also true:

The greater the quantity added, the greater is the result.

For example:

$$-7 \text{ is greater than } -10$$

$$3+(-7) \;\ldots\ldots\ldots\ldots\; 3+(-10)$$

for we see that $-4 \;\ldots\ldots\ldots\ldots\; -7$

Similarly, the *less the quantity subtracted, the greater is the result.* From 3 subtract -8, the result is $+11$; subtract less than -8, say -12, the result is $+15$, greater than $+11$. From -4 subtract 7, the result is -11; subtract less than 7, say -3, and the result is $-4-(-3)$, or -1, greater than -11. And it will be found that all such theorems relative to addition or subtraction as are true of arithmetical, will also be true of algebraical, quantities; which is the particular advantage of our new definition. For instance, remark the following:

If a be greater than b, $a-b$ is positive; if a be less than b, $a-b$ is negative. Thus

$$-3-(-4) = +1 \qquad -3-(-2) = -1$$

The signs of greater and less are $>$ and $<$. Thus,

For a is greater than b, write $a > b$.

For a is less than b, write $a < b$.

The angle is turned towards the greater quantity. In the sign of equality $=$ there is no angle towards either quantity.

---

arithmetical meaning, except when arithmetical quantities are mentioned. Those who object to it in its present shape, will of course object to the zoological supposition in page 58.

5. *Multiplication* and *Division*. These rules are, so far as the numerical quantities are concerned, the same as in arithmetic. The rule of signs, as we have seen, is, *like* signs produce +, *unlike* signs —.

$$+ab \text{ is both } +a\times\ +b \text{ and } -a\times\ -b$$

$$-ab \text{ is both } -a\times\ +b \text{ and } +a\times\ -b$$

$$+\frac{a}{b} \text{ is both } \frac{+a}{+b} \text{ and } \frac{-a}{-b}$$

$$-\frac{a}{b} \text{ is both } \frac{-a}{+b} \text{ and } \frac{+a}{-b}$$

The terms greater and less cannot always be applied to products as in arithmetic. For instance, $3>2$, $5>4$, and therefore, $3\times 5>2\times 4$; but from $3>-2$, $-3>-4$, it does not follow that $3\times -3>-2\times -4$, or $-9>8$, but the contrary $-9<8$. But it is seldom necessary to deduce the algebraical magnitude of a product from that of its factors; we therefore leave to the student the collection of the different cases.

6. *Proportion*. Four quantities are said to be proportional when the first divided by the second is equal to the third divided by the fourth. This definition is the same in words as the definition of proportion in arithmetic, but the words *quantity*, *divided by*, and *equal*, have their extended signification. The words greater and less cannot always be applied as in arithmetic. Thus $\frac{3}{-4}$ being $\frac{-6}{8}$ we have $3:-4::-6:8$, where 3 is greater than -4, but -6 is less than 8.

We shall now apply our definitions to a problem, and shall choose the cases already noticed in page 45, as being two different problems when only equations of the first degree are to be used.

A has £60, and is to receive the absolute balance that appears in B's books, whether for or against B; but C, who has £200, is to take B's property and pay his debts. After doing this it is found that C's property is 3 times that of A. What is the absolute balance for or against B?

Let x be this balance, positive or negative according as it is for or against B; then A has $60\pm x$, the positive sign being used* when x

* For A's property is to be increased on either supposition; hence, if the balance be $+3$, he must have $60+(+3)$; if it be -3, he must have $60-(-3)$.

is positive, the negative when x is negative. But C has $200 + x$; therefore,

$$3(60 \pm x) = 200 + x$$

or

$$\pm 3x = 20 + x$$

This contains two equations, one for the positive, one for the negative sign. But from it follows that

$$\pm 3x \times \pm 3x = (20 + x)(20 + x)$$

in which the first side is $+9xx$ in both cases, for $-3x \times -3x$ and $+3x \times +3x$, are the same, namely, $+9xx$. Therefore,

$$+9xx = 400 + 40x + xx$$

or

$$8xx - 40x - 400 = 0$$

$(\div)8$

$$xx - 5x - 50 = 0$$

When we come to the solution of equations of the second degree, we shall find that this equation can be true only for two values of x; either $x = 10$, or $x = -5$. That is, the balance is either £10 for B, or £5 against him; which are the solutions already found in page 45.

That either 10 or -5 will satisfy the preceding, may be shewn as follows:

$x = 10$	$x = -5$
$xx = 100$	$xx = 25$
$-5x = -50$	$-5x = +25$
$-50 = -50$	$-50 = -50$
$xx - 5x - 50 = 0$	$xx - 5x - 50 = 0$

Since the extensions of algebra have been so laid down that the rules for managing algebraical quantities when they are not arithmetical are the same as those which must be employed when they are arithmetical, it follows that the arithmetical case of a problem may be taken as a guide; for, to say that certain operations follow the same rules as in arithmetic, or that we must proceed as if the operation was arithmetical, is only the same thing in different words.

Up to page 56 we have considered the symbols of algebra, which are not arithmetical, as *results of misconception*, and have called the rules by which they are treated *corrections*. In page 59, &c., by properly laying down definitions, these same symbols are recognised and expected, so that the term erroneous no longer applies to them. The student should not immediately give over the first method of considering them, but should frequently, while employed upon the rules (pages 49, &c.), make himself sure that he understands the connexion of the preceding method with those rules; and in future we may accordingly employ both methods, it being always understood that when the first is used, the extensions required by the second are dropped for the moment.

The following examples of the use of the rules are added for practice. Such symbols as $\bar{a}$ and $-a$ are used indiscriminately, it being remembered that they mean the same thing in practice, but are referred to two different methods of considering the subject in theory.

$$\bar{8} \times 4 \div \bar{3} = 10\tfrac{2}{3} \qquad \bar{8} \times \bar{4} \div \bar{3} = -10\tfrac{2}{3} \text{ or } \overline{10\tfrac{2}{3}}$$

$$\bar{6} + \bar{4} - \overline{12} = 2 \qquad (-6) + (-8) + (-13) = -27$$

$$\frac{\bar{a}\,\bar{b} + c\,\bar{d}}{\bar{m}} = \frac{cd - ab}{m} \qquad \frac{(-a) \times (-b) \times c}{-d} = -\frac{abc}{d}$$

$$\frac{a - \bar{b}}{a + \bar{b}} \times \bar{c} = \frac{a + b}{b - a}\,c \qquad \frac{-a - (-b)}{a + (-b)} = -1$$

$$a\bar{b} - b\bar{a} = 0 \qquad a(-b) + b(-a) = -2ab$$

$$a\bar{b}c = \bar{a}bc = ab\bar{c} = \bar{a}\,\bar{b}\,\bar{c} = -abc$$

CHAPTER III.

EQUATIONS OF THE FIRST DEGREE WITH MORE THAN ONE UNKNOWN QUANTITY.

Let there be such an equation as

$$x+y = 12$$

resulting from a problem which involves two unknown quantities, x and y. Such a one is the following:

Problem. A is a given point in a straight line, and B and C are two other points. From A to half-way between B and C it is six feet. What are the distances of B and C from A?

Let us take as our principal case that in which B and C are both on the same side of A. Let $AB = x$ feet, $AC = y$ feet; then $BC = x-y$, and from B to half-way between B and C the distance is $\frac{1}{2}(x-y)$; whence from A to the same middle point the distance is

$$x - \frac{1}{2}\left(x-y\right) \text{ which is therefore } = 6$$

$$(\times)2 \qquad 2x-(x-y) = 12 \text{ or } x+y = 12$$

Here the problem is what is called *indeterminate*, that is, admitting of an infinite number of solutions. All that is laid down relative to x and y is found to do no more than require that their sum shall be 12, which can be satisfied in an infinite number of ways; for all the following cases are solutions, and others might be made at pleasure.

$$x = 1 \quad y = 11 \qquad x = \frac{1}{2} \quad y = 11\frac{1}{2}$$

$$x = 2 \quad y = 10 \qquad x = 2\frac{1}{4} \quad y = 9\frac{3}{4}$$

$$x = 3 \quad y = 9 \qquad x = 3\frac{2}{5} \quad y = 8\frac{3}{5}$$

&c. &c. &c. &c.

Again, when either x or y is taken negatively, we have solutions such as the following:

$$x = -1 \quad y = +13 \qquad\qquad x = 15 \quad y = -3$$
$$x = -2 \quad y = +14 \qquad\qquad x = 16 \quad y = -4$$
$$x = -1\tfrac{1}{2} \quad y = +13\tfrac{1}{2} \qquad\qquad x = 12\tfrac{3}{5} \quad y = -\tfrac{3}{5}$$

&c. &c. &c. &c.

the first column corresponding, with the explanations before given at pages 14–19, to the case in which B only is at the left of A; and the second to the case in which C only lies to the left of A. Thus, if B be 1 foot to the left of A, and C 13 feet to the right, the middle point between C and B is at 6 feet distance from A to the right. Observe that we cannot in this equation of the first degree, $x+y=12$, include the cases in which the middle point falls to the left of A; for since this quantity 6, given in the problem, has been treated throughout as an arithmetical (or algebraical positive, see page 59) quantity, the equation formed from it cannot include those cases of the problem in which the corresponding line is so measured that its symbol ought to be negative.*

From the preceding and similar cases, we deduce the following principles:

1. One equation between two unknown quantities admits of an infinite number of solutions; either of the unknown quantities may be what we please, and the equation can be satisfied by giving a proper value to the other.

2. A problem which gives rise to such an equation is indeterminate, or admitting of an indefinite number of solutions.

Let us now suppose two equations, each containing the same two unknown quantities. For instance,

$$x+y = 12 \qquad\qquad 3x-2y = 31$$

* On reading the problem again, therefore, we perceive either that we have not sufficiently defined it, by omitting to state whether the six feet is to the right or the left of A, or else that there are two problems involved in it, as in page 45, or that these two must be represented in one equation of the second degree, as in page 64. For the student who is disposed to try to represent the two cases in one equation, we give the result, namely,

$$xx+yy+2xy = 144.$$

the first (considered by itself) has an infinite number of solutions; so also has the second. As follows:

Solutions of the first equation.		Solutions of the second equation.	
$x = 10$	$y = 2$	$x = 10$	$y = -\frac{1}{2}$
$x = 10\frac{1}{2}$	$y = 1\frac{1}{2}$	$x = 10\frac{1}{2}$	$y = \frac{1}{4}$
$x = 11$	$y = 1$	$x = 11$	$y = 1$
$x = 11\frac{1}{4}$	$y = \frac{3}{4}$	$x = 11\frac{1}{4}$	$y = \frac{11}{8}$
&c.	&c.	&c.	&c.

We have taken the same set of values for x in both; and we find the corresponding values of y different, generally speaking, but the same in one particular case: that is, we find a set of values $x = 11$, $y = 1$, which satisfies both equations. The question now is, among all the infinite number of sets of values which satisfy one or the other equation, how many are there which satisfy both? There is only one, as we shall find from the following process of solution.

If $x + y = 12$, it follows that $x = 12 - y$. Substitute this value of x in the second equation (which may be done, since the solutions of the second which we wish to obtain are only those which are also solutions of the first), and we find $3(12 - y) - 2y = 31$, or $36 - 5y = 31$, or $y = 1$. It appears then, that the supposition of both equations being true at once consists with no other value of y except 1, or (since $x + y = 12$) of x except 11.

Let the equations proposed be

$$ax + by = c \qquad\qquad px + qy = r$$

where a, b, c, p, q, and r, are certain known quantities.

First Method. Obtain a value of one of the unknown quantities from one equation and substitute it in the other. The resulting equation will then have only one unknown quantity.

From the first equation,

$$x = \frac{c - by}{a}$$

which value substituted in the second gives

$$\frac{pc - pby}{a} + qy = r \qquad y = \frac{ar - cp}{aq - bp}$$

To obtain x, find y from the first equation, and repeat the process, which gives

$$y = \frac{c - ax}{b}, \quad px + \frac{qc - qax}{b} = r, \quad x = \frac{cq - br}{aq - bp}.$$

Or substitute the value of y first obtained in the previous expression for x; thus,

$$x = \frac{c - by}{a}, \qquad by = \frac{bar - bcp}{aq - bp}$$

$$c - by = \frac{caq - cbp - (bar - bcp)}{aq - bp} = \frac{caq - bar}{aq - bp}$$

$$= \frac{a(cq - br)}{aq - bp} \therefore x \text{ or } \frac{c - by}{a} = \frac{cq - br}{aq - bp}.$$

Verification. If $x = \dfrac{cq - br}{aq - bp}$, and $y = \dfrac{ar - cp}{aq - bp}$

$$ax + by = \frac{acq - abr}{aq - bp} + \frac{abr - bcp}{aq - bp}$$

$$= \frac{acq - bcp}{aq - bp} = \frac{c(aq - bp)}{aq - bp} = c$$

$$px + qy = \frac{cpq - bpr}{aq - bp} + \frac{aqr - cpq}{aq - bp} = r$$

Second Method. Multiply both the equations in such a way that the terms which contain the same unknown quantities may have the same co-efficient; then add or subtract the two results, whichever will cause the similar terms to disappear. Generally, the shortest method is: Multiply each equation by the co-efficient which the quantity not wanted has in the other equation.

To find y.

$$ax + by = c \qquad (\times)p \quad pax + pby = pc$$

$$px + qy = r \qquad (\times)a \quad pax + qay = ar$$

$$(-) \qquad aqy - bpy = ar - cp$$

$$(\div)\overline{aq - bp} \qquad y = \frac{ar - cp}{aq - bp}$$

To find x.

$$ax + by = c \qquad (\times)q \quad aqx + bqy = qc$$

$$px + qy = r \qquad (\times)b \quad bpx + bqy = br$$

$$(-) \qquad aqx - bpx = cq - br$$

$$(\div)\overline{aq - bp} \qquad x = \frac{cq - br}{aq - bp}$$

Third Method. Obtain a value of one of the unknown quantities from each of the equations, and equate the values so obtained.

From the first equation $y = \frac{c - ax}{b}$ $x = \frac{c - by}{a}$

From the second equation $y = \frac{r - px}{q}$ $x = \frac{r - qy}{p}$

$$\therefore \quad \frac{c - ax}{b} = \frac{r - px}{q} \qquad x = \frac{cq - br}{aq - bp}$$

$$\frac{c - by}{a} = \frac{r - qy}{p} \qquad y = \frac{ar - cp}{aq - bp}$$

Let the student now repeat all the three processes with the following equations (see page 38).

$$\left.\begin{array}{r} ax + by = c \\ a'x + b'y = c' \end{array}\right\} \text{ which give } \left\{\begin{array}{l} x = \dfrac{cb' - bc'}{ab' - ba'} \\ y = \dfrac{ac' - ca'}{ab' - ba'} \end{array}\right.$$

(1.) $3x - 2y = 14$ $\quad (\times)2 \quad 6x - 4y = 28$

$2x + 3y = 100$ $\quad (\times)3 \quad 6x + 9y = 300$

$(-)$ $13y = 272$ $\quad y = 20\frac{12}{13}$

$(\times)3$ $9x - 6y = 42$ $\quad (\times)2 \quad 4x + 6y = 200$

$(+)$ $13x = 242$ $\quad x = 18\frac{8}{13}$

(2.) $x + y = a$ $\quad x - y = b$

$(+)$ $2x = a + b$ $\quad x = \frac{a+b}{2}$

$(-)$ $2y = a - b$ $\quad y = \frac{a-b}{2}$

(3.) $px + y = 1$ $\quad x - py = 2$

The first is $px + y = 1$

The second, $(\times)p$ $\quad px - ppy = 2p$

$(-)$ $\quad y + ppy = 1 - 2p$ $\quad y = \frac{1 - 2p}{1 + pp}$

$$1 - y = \frac{pp + 2p}{1 + pp} \qquad x = \frac{1 - y}{p} = \frac{p + 2}{1 + pp}$$

(4.) $3x-7 = 4+(x+y)$ or $2x-y = 11$

$2y+79 = 5x$ or $5x-2y = 79$

The first, $(\times)5 \quad 10x-5y = 55$

The second, $(\times)2 \quad 10x-4y = 158$

$(-) \quad y = 103$

From the first, $x = \dfrac{11+y}{2} = 57$

(5.) $3x+4y = 13 \qquad 4x+5y = 10$

$(\times)4 \quad 12x+16y = 52 \qquad (\times)3 \quad 12x+15y = 30$

$(-) \quad y = 52-30 = 22$

$(\times)5 \quad 15x+20y = 65 \qquad (\times)4 \quad 16x+20y = 40$

$(-) \quad x = 40-65 = -25$

the problem producing these equations must therefore be treated as before described.

Having given $\begin{matrix} ax+by = c \\ px+qy = r \end{matrix} \quad x = \dfrac{cq-br}{aq-bp} \quad y = \dfrac{ar-cp}{aq-bp}$

we may find the following:

$$\left\{\begin{matrix} ax-by = c \\ px-qy = r \end{matrix}\right. \quad x = \frac{cq-br}{aq-bp} \quad y = \left.\frac{cp-ar}{aq-bp}\right\}$$

$$\left\{\begin{matrix} ax-by = c \\ qy-px = r \end{matrix}\right. \quad x = \frac{cq+br}{aq-bp} \quad y = \left.\frac{ar+cp}{aq-bp}\right\}$$

On looking at the two first sets of equations, we see that they differ in this, that $+by$ and $+qy$ in the first are replaced by $-by$ and $-qy$ in the second. But we know that $ax-by$ is the corrected form of $ax+\bar{b}y$ (see page 46, &c.), and $px-qy$ of $px+\bar{q}y$; and since we have proved that the corrections may be deferred to any stage of the process, it follows that we may take the solutions of

$$\begin{matrix} ax+\bar{b}y = c \\ px+\bar{q}y = r \end{matrix} \quad x = \frac{c\bar{q}-\bar{b}r}{a\bar{q}-\bar{b}p} \quad y = \frac{ar-cp}{a\bar{q}-\bar{b}p}$$

and conclude, that the corrected solutions will be the true result of the corrected equations. This if done will give results as above; for the value of x, corrected by the rules in page 52, gives

$$\frac{-cq+br}{-aq+bp} \text{ or } \frac{br-cq}{bp-aq} \text{ or } \frac{cq-br}{aq-bp} \text{ (Page 64).}$$

and a similar process must be followed for y.* In this way, any results derived from an expression such as $ax+by+cz$, may be made to furnish the corresponding results which would have been derived from $ax-by-cz$, $cz+by-ax$, or any other variation which is only *in sign*. We write underneath the different cases which may arise, and the manner of referring them to the one which is chosen as the representative of all.

The following	May be considered as	Or as
$ax+by-cz$	$ax+by+\bar{c}z$	$ax+by+c\bar{z}$
$ax-by-cz$	$ax+\bar{b}y+\bar{c}z$	$ax+b\bar{y}+c\bar{z}$, &c.
$by-ax-cz$	$\bar{a}x+by+\bar{c}z$	$\bar{a}x+by+c\bar{z}$, &c.
&c.	&c.	&c.

Any other case may be taken as the representative of the whole: thus, if it were $ax-by-cz$, then $ax+by+cz$ must be considered as $ax-\bar{b}y-\bar{c}z$, &c.

PROBLEM.† *On the intersection of straight lines.*

Principle. It is proved in geometry, that if any angle BOA (for simplicity, we use a right angle) be drawn, and a straight line AB cutting the sides which contain it in A and B, and if from P, any point *between A and B*‡ parallels be drawn, PN to OA, and PM to OB, and if OA, OB, PM, and PN be measured in inches, or tenths of inches, or any other convenient unit (provided it be the same for all); then (PM meaning, not the line itself, but the number of units in it) the product of PM and OA, added to the product of PN and OB, will be equal to the product of OA and OB, or

$$PM\times OA+PN\times OB=OA\times OB$$

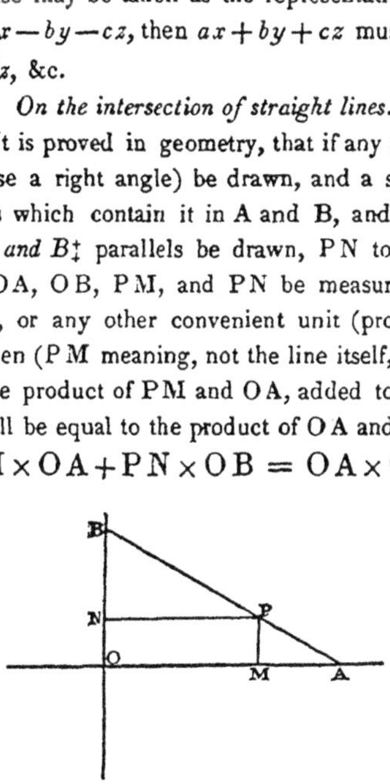

* Attend here to the remarks in the Preface, on the necessity of working more examples than are given in the book.

† If the beginner have no knowledge of the most common terms of geometry, he must either acquire it, or omit this problem altogether.

‡ From this supposition we start; we shall afterwards make use of

Let there be two such lines, AB and A′B′ (draw a large figure and insert A′B′), *cutting one another in the angle* BOA in P; having given OA, OB, OA′, and OB′, required the value of PM and PN.

Let $\quad OA = 10$ units $\qquad OA' = 7$ units
$\qquad OB = 8$ .. $\qquad OB' = 15$ ..

$$PN = x \text{ units}$$
$$PM = y \quad ..$$

Then, because P is on the line AB (preceding principle),

$$OA \times PM + OB \times PN = OA \times OB \quad \text{or} \quad 10y + 8x = 80$$

Again, because P is also on A′B′ (similar reason),

$$OA' \times PM + OB' \times PN = OA' \times OB' \quad \text{or} \quad 7y + 15x = 105$$

Here then are two equations which y and x must satisfy. Solving them by either of the preceding methods, we have

x or PN is $5\frac{10}{47}$ units; y or PM is $3\frac{39}{47}$ units.

General Case. Let $OA = a$ units $\qquad OA' = a'$ units.
$\qquad OB = b$ .. $\qquad OB' = b'$ ..

$$PN = x \text{ units}$$
$$PM = y \quad ..$$

The equations to be solved, which are followed by the solution at length, as at page 70, will then be

$$ay + bx = ab$$
$$a'y + b'x = a'b'$$

$$aa'y + a'bx = aa'b \qquad ab'y + bb'x = abb'$$
$$aa'y + ab'x = aa'b' \qquad a'by + bb'x = a'bb'$$
$$(a'b - ab')x = aa'(b - b') \qquad (a'b - ab')y = bb'(a' - a)$$
$$x = aa'\frac{b - b'}{a'b - ab'} \qquad y = bb'\frac{a' - a}{a'b - ab'}$$

---

the preceding theory of algebraical symbols to extend our results to the case where P is *not* between A and B. All words in *italics* point out what is peculiar to the first case.

Verification.

$$ay + bx = abb'\frac{a'-a}{a'b-ab'} + aba'\frac{b-b'}{a'b-ab'}$$

$$= ab\left\{\frac{a'b'-ab'}{a'b-ab'} + \frac{a'b-a'b'}{a'b-ab'}\right\}$$

$$= ab\frac{a'b-ab'}{a'b-ab'} = ab$$

$$a'y + b'x = a'bb'\frac{a'-a}{a'b-ab'} + aa'b'\frac{b-b'}{a'b-ab'}$$

$$= a'b'\left\{\frac{a'b-ab}{a'b-ab'} + \frac{ab-ab'}{a'b-ab'}\right\}$$

$$= a'b'\frac{a'b-ab'}{a'b-ab'} = a'b'$$

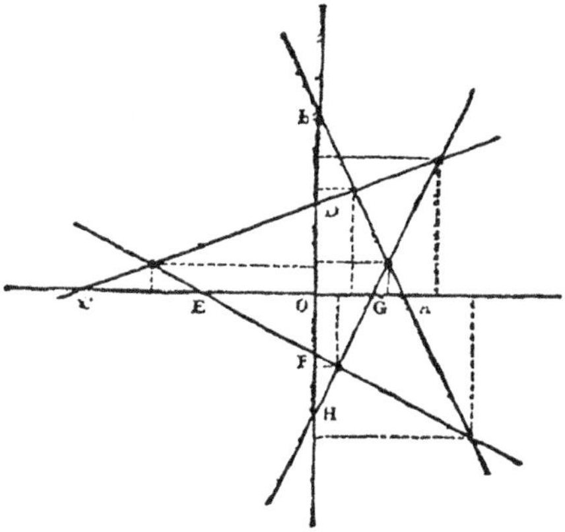

In the preceding figure are four lines, AB, CD, EF, and GH, cutting the axes* in as many different ways as is possible, that is, no two intersecting *both* OA and OB on the same side of O. There are therefore, six points of intersection (as in the figure), and the perpendiculars drawn from them to the axes OA and OB are inserted; but no letters are put to them, it being understood that P is

* Axes, the principal lines OA and OB containing the angle first mentioned, and lengthened in both directions.

the point of intersection under consideration,* while PM and PN are the perpendiculars, as in the first figure.

Let the distances at which the four lines cut the axes be as follows, in numbers of the unit chosen:

$$\begin{array}{llll} OA = 3 & OC = 8 & OE = 4 & OG = 2 \\ OB = 6 & OD = 3 & OF = 2 & OH = 4 \end{array}$$

Let us first take the intersection of A B and C D. Place the point P at this intersection, and let P M $= y$ and P N $= x$, as in the first figure. Then, because P is on the line A B, we have, from the principle stated at the outset,

$$3y + 6x = 18$$

But on looking at the line C D, we observe, 1. That the principle does not apply directly to it, because C D is not contained in the angle B O A, but in the contiguous angle B O C. 2. That O C (in length 8 units) is not measured in the direction O A, but in the contrary direction. If, therefore, we write $\bar{8}$ instead of 8, and with this form the same equation as we should have formed if O C had been measured towards A, we shall then, by correcting the equation as in page 66, obtain another equation with which to proceed.† That is, because the point P is on C D, we have

$$\bar{8}y + 3x = \bar{8} \times 3 \text{ or (page 66) } -8y + 3x = \overline{24} = -24$$
$$\text{or} \quad 8y - 3x = 24$$

[For, since $3x - 8y = -24$ denotes that $8y$ exceeds $3x$ by 24, this may be written $8y - 3x = 24$.]

Solving the two equations thus obtained, namely,

$$\left.\begin{array}{r} 3y + 6x = 18 \\ 8y - 3x = 24 \end{array}\right\} \text{ we get } \left\{\begin{array}{l} x = 1\frac{5}{19} \\ y = 3\frac{9}{19} \end{array}\right.$$

which, with the succeeding results, may be verified by measurement, as well as from the equations.

* Let the student draw this figure on a large scale in ink, and mark the letters belonging to the point under consideration in pencil, to be rubbed out on passing to a new point.

† We cannot refer to any single page here. This principle is the total result of Chapter II.

To apply the same method to every case, we must distinguish all the lines O A, O B, &c. by the negative sign, which do not fall in the directions of O A or O B. That is,

$$OA = 3 \quad OC = \bar{8} \quad OE = \bar{4} \quad OG = 2$$
$$OB = 6 \quad OD = 3 \quad OF = \bar{2} \quad OH = \bar{4}$$

Proceeding with these as in the last case, we have as follows:

Intersection of	Equations.	Corrected equations.
1. A B and C D	$\begin{cases} 3y+6x=18 \\ \bar{8}y+3x=\overline{24} \end{cases}$	$\begin{cases} 3y+6x=18 \\ 8y-3x=24 \end{cases}$
2. A B and E F	$\begin{cases} 3y+6x=18 \\ \bar{4}y+\bar{2}x=\bar{4}\times\bar{2} \end{cases}$	$\begin{cases} 3y+6x=18 \\ 4y+2x=-8 \end{cases}$
3. A B and G H	$\begin{cases} 3y+6x=18 \\ 2y+\bar{4}x=2\times\bar{4} \end{cases}$	$\begin{cases} 3y+6x=18 \\ 2y-4x=-8 \end{cases}$
4. C D and E F	$\begin{cases} \bar{8}y+3x=\overline{24} \\ \bar{4}y+\bar{2}x=\bar{4}\times\bar{2} \end{cases}$	$\begin{cases} 8y-3x=24 \\ 4y+2x=-8 \end{cases}$
5. C D and G H	$\begin{cases} \bar{8}y+3x=\overline{24} \\ 2y+\bar{4}x=2\times\bar{4} \end{cases}$	$\begin{cases} 8y-3x=24 \\ 2y-4x=-8 \end{cases}$
6. E F and G H	$\begin{cases} \bar{4}y+\bar{2}x=\bar{4}\times\bar{2} \\ 2y+\bar{4}x=2\times\bar{4} \end{cases}$	$\begin{cases} 4y+2x=-8 \\ 2y-4x=-8 \end{cases}$

These corrected equations are some of them not arithmetically true; for instance, $4y+2x=-8$. But remember that they are made on the supposition that P M (y) and P N (x) are measured in such a direction that P falls within the angle A O B; which may not be the case. We may, however, apply the rules in page 70 to their solution; of which we shall give one case (the second) complete, as follows:

$$3y + 6x = 18$$
$$4y + 2x = -8$$

Multiply the second equation by 3.

$$12y + 6x = -8 \times 3 \quad \text{or} \quad 12y + 6x = -24$$

Subtract this from the first, which gives

$$(3-12)y = 18-(-24) = 18+24 = 42$$

$$y = \frac{42}{3-12} = \frac{42}{-9} = -\frac{42}{9} = -4\frac{2}{3}$$

From the first equation,

$$x = \frac{18-3y}{6} = \frac{18-3(-4\frac{2}{3})}{6} = \frac{18+14}{6} = 5\frac{1}{3}$$

On placing the point P at the intersection of AB and EF, we see that PM (y) is measured in a *contrary* direction to the one supposed, and PN (x) in the same direction, as would be presumed from the negative sign of the first, and the positive sign of the second.

Proceeding in the same way, we have the following values of x and y in the six cases :

Intersection of,	Value of x (PN).	Value of y (PM).
1. AB and CD	$1\frac{5}{19}$	$3\frac{9}{19}$
2. AB and EF	$5\frac{1}{3}$	$-4\frac{2}{3}$
3. AB and GH	$2\frac{1}{2}$	1
4. CD and EF	$-5\frac{5}{7}$	$\frac{6}{7}$
5. CD and GH	$4\frac{4}{13}$	$4\frac{8}{13}$
6. EF and GH	$\frac{4}{5}$	$-2\frac{2}{5}$

[Remember that in such an expression as $-1\frac{1}{2}$, the $-$ refers to the whole ; that is, it is $-(1+\frac{1}{2})$ or $-1-\frac{1}{2}$, not $-1+\frac{1}{2}$.]

Upon examining the six lines by a well-constructed figure, it will appear that whenever an answer is negative, it is measured in a direction contrary to that which was supposed in the principle at the beginning.

The corrections might have been deferred to the end of the process, in the following manner :

The equations (see page 74)

$$ay + bx = ab \qquad a'y + b'x = a'b'$$

give

$$x = aa'\frac{b-b'}{a'b-ab'} \qquad y = bb'\frac{a'-a}{a'b-ab'}$$

Take the two following sets (the second being case 6),

$$ay + bx = ab \qquad \bar{4}y + \bar{2}x = \bar{4} \times \bar{2}$$
$$a'y + b'x = a'b' \qquad 2y + \bar{4}x = 2 \times \bar{4}$$

which agree if

$$a = \bar{4} \qquad b = \bar{2} \qquad a' = 2 \qquad b' = \bar{4}$$

Substitute these values in the expressions for x and y, which give,

$$x = \bar{4} \times 2 \times \frac{\bar{2} - \bar{4}}{2 \times \bar{2} - \bar{4} \times \bar{4}} = (-8) \times \frac{-2+4}{-4-16}$$
$$= (-8) \times \left(-\frac{2}{20}\right) = \frac{16}{20} = \frac{4}{5}$$
$$y = \bar{2} \times \bar{4} \times \frac{2 - \bar{4}}{2 \times \bar{2} - \bar{4} \times \bar{4}} = 8 \times -\frac{6}{20} = -2\frac{2}{5}$$

the same as before. We leave this problem, and proceed.

We have shewn, page 67, that one equation between two unknown quantities admits of an infinite number of solutions; and that a second equation must be given by the problem, or it is not reducible to a single answer. But this second equation must be *independent* of the first, that is, must not be one of those which can be reduced to the first. For instance, $x + y = 12$ admits of an infinite number of solutions; and if the second equation be either of the following,

$$2x + 2y = 24 \qquad 3x + 3y = 36 \qquad \frac{1}{2}x + \frac{1}{2}y = 6$$
$$3x - 18 = 18 - 3y \qquad 2x + y = 24 - y, \text{ \&c.}$$

the same infinite number of solutions will still exist; for if $x + y = 12$, all the equations just given must be true. That is, instead of giving two equations, we have only given the same equation in two different forms.

Now, we have already found (page 25), that in one case the index of an infinite number of results was the appearance of the result in the form $\frac{0}{0}$. We proceed to see whether this will be the case here.

$$\text{If } \left\{\begin{aligned} ax + by &= c \\ px + qy &= r \end{aligned}\right\} \text{ then } x = \frac{cq - br}{aq - bp} \qquad y = \frac{ar - cp}{aq - bp}$$

To try a case in which the second equation is dependent on the first, suppose

$$p = ma \qquad q = mb \qquad r = mc$$

in which case the second equation becomes

$$max+mby=mc \qquad (\div)m \quad ax+by=c$$

the same as the first. Substitute the above values of p, q, and r, in the values of x and y, which gives

$$x=\frac{cmb-bmc}{amb-bma}=\frac{0}{0} \qquad y=\frac{amc-cma}{amb-bma}=\frac{0}{0}$$

in which the same anomaly appears as in page 25, and with the same interpretation.

If there be *three* unknown quantities, by reasoning similar to that in page 67, we may shew that there must be as many as *three* independent equations, or else the problem admits an infinite number of solutions. We will shew in one instance the method of proceeding.

Let

$$2x+4y-3z=10 \;\cdots\cdots\; (1)$$
$$5x-3y+2z=20 \;\cdots\cdots\; (2)$$
$$3x+2y+5z=50 \;\cdots\cdots\; (3)$$

Multiply both sides of (1) by 2, and of (2) by 3, in the results of both of which z will have the same coefficient.

$$\begin{array}{ll} \text{Equ. (1)} \times 2 & 4x+8y-6z=20 \\ \text{Equ. (2)} \times 3 & 15x-9y+6z=60 \\ \hline (+) & 19x-y=80 \;\cdots\cdots\cdots\; (4) \end{array}$$

Repeat a similar process with equations (2) and (3).

$$\begin{array}{ll} \text{Equ. (2)} \times 5 & 25x-15y+10z=100 \\ \text{Equ. (3)} \times 2 & 6x+4y+10z=100 \\ \hline (-) & 19x-19y=0 \\ (\div)19 & x-y=0 \text{ or } x=y \;\cdots\cdots\; (5) \end{array}$$

Two equations are thus found (4) and (5), containing x and y only, not z. These solved, give

$$y=\frac{40}{9} \qquad x=\frac{40}{9}$$

Substitute these values in either of the three given equations, the second, for example. Then

$$5\times\frac{40}{9}-3\times\frac{40}{9}+2z=20 \qquad z=\frac{50}{9}$$

There is an artifice* which is useful when one only of the three unknown quantities is required. Suppose, for example, that in the preceding equations only the value of z is wanted. Take two new quantities, m and n, not yet known, but afterwards to be determined in any manner that may be convenient.† Since the two sides of an equation may be multiplied by any quantity, multiply (2) by m, and (3) by n, and add the results to (1). This will give

$$(2+5m+3n)x+(4-3m+2n)y$$
$$+(2m+5n-3)z=10+20m+50n \cdot\cdot (A)$$

Since m and n may be taken at pleasure, and since the value of z only is wanted, let m and n be such that

$$2+5m+3n=0 \quad \text{or} \quad 5m+3n=-2$$
$$4-3m+2n=0 \quad \text{or} \quad 3m-2n=4$$

then m and n must be the solutions of the preceding equations, which give $m=\frac{8}{19}$, $n=-\frac{26}{19}$. But in (A) the terms which contain x and y disappear, being multiplied by 0. Therefore,

$$(2m+5n-3)z = 10+20m+50n$$

$$z=\frac{10+20m+50n}{2m+5n-3}=\frac{10+20\times\frac{8}{19}+50\left(-\frac{26}{19}\right)}{2\times\frac{8}{19}+5\left(-\frac{26}{19}\right)-3}$$

(Multiply numerator and denominator of this fraction by 19, and reduce it),

$$=\frac{10\times 19+20\times 8-50\times 26}{2\times 8-5\times 26-3\times 19}=\frac{-950}{-171}=\frac{950}{171}=\frac{50}{9}$$

Anomalies. A problem may give rise to two equations which are absolutely incompatible with each other, such as the following:

$$x+y=12 \qquad\qquad x+y=13$$

or it may happen, that if there be three unknown quantities and three equations, one of the latter may be impossible if *both* the others be true, though it be not inconsistent with either of the others *singly*. For instance, take

$$x-y=10 \qquad y-z=11 \qquad x-z=12$$

* An artifice is a name given to any process by which either the principle or practice of any method is shortened, either generally, or in any particular case.

† Such quantities are usually called *arbitrary*.

The 1st and 2d are true if	$x = 20$	$y = 10$	$z = -1$
The 1st and 3d	$x = 20$	$y = 10$	$z = 8$
The 2d and 3d	$x = 20$	$y = 19$	$z = 8$

But no values of x, y, and z can satisfy all three together, as will be evident by adding the first two. For if $x - y = 10$, and $y - z = 11$, then

$$(x - y) + (y - z) = 21 \quad \text{or} \quad x - z = 21$$

which is inconsistent with the third equation.

It is left to the student to examine the general solution in page 70, and to shew that when the two equations become incompatible, the values of x and y take the form discussed in page 21, namely, that of a fraction which has 0 for its denominator. It might also be shewn, that the problems which give rise to incompatible equations admit of an interpretation similar to that already derived from results of the form $\frac{c}{0}$ in the page last quoted.

It will be found, that two equations such as are treated in this chapter will in no case become inconsistent except when they can be so reduced as to have their first sides identical, and their second sides different; such as

$$2x + 3y = 10 \qquad\qquad 2x + 3y = 12$$

The subtraction of the first from the second would give $0 = 2$, an absurdity of the same character as $ax = ax + c$ (page 23), from which the form $\frac{c}{0}$ was first derived.

CHAPTER IV.

ON EXPONENTS, AND ON THE *CONTINUITY* OF ALGEBRAIC EXPRESSIONS.

THE continual occurrence of the multiplication of the same quantity two, three, or more times by itself, has rendered some abbreviations necessary, which we proceed to explain.

x multiplied by x, or xx, is called the second power of x.
xx .... x, or xxx, ... the third power of x.
xxx .... x, or $xxxx$, ... the fourth power of x.

and so on. Or, n $x$es* multiplied together give what is called the nth power of x. The second and third powers are usually called the *square* and *cube*. Thus xx is called the square of x, and is read *x square;* and xxx is called the cube of x, and is read *x cube:* and a number multiplied by itself is said to be squared; multiplied twice by itself, it is said to be cubed, &c.

By an extension, x itself is called the first power of x.

The abbreviated representation of a power is as follows: over the letter raised to any power, on the right, place the number of times the letter is seen in that power. Thus,

xx is written x^2
xxx .... x^3
$xxxx$ .... x^4

and so on. Here 2, 3, 4, &c. are called *exponents* of x. Similarly,

$(a+b)\times(a+b)$ is written $(a+b)^2$
$(a+b)\times(a+b)\times(a+b)$ .... $(a+b)^3$

and so on. The following results may now be easily found:

* The beginner's common mistake is, *that* x *multiplied* n *times by* x *is the* n*th power.* This is not correct; x multiplied *once* by x (xx) is the *second* power; x multiplied n times by x is the $(n+1)$th power.

$$x \times x = x^2 \qquad x^2 \times x = x^3 \qquad x^3 \times x = x^4, \text{ \&c.}$$

$$(a+x)^2 = a^2+2ax+x^2$$
$$(a-x)^2 = a^2-2ax+x^2$$
$$(a+x)(a-x) = a^2-x^2$$
$$(a^2+ax+x^2)(a-x) = a^3-x^3$$
$$(a+b)^3 = a^3+3a^2b+3ab^2+b^3$$
$$(a-b)^3 = a^3-3a^2b+3ab^2-b^3$$

To multiply together any two powers of the same letter, let the exponent of the product be the sum of the exponents of the factors. Thus, to multiply x^3 and x^4,

$$x^3 \text{ is } xxx$$
$$x^4 \text{ is } xxxx$$
$$\therefore \quad x^3 \times x^4 \text{ is } xxxxxxx \text{ or } x^7 \text{ or } x^{3+4}$$

By the extension previously made of calling x the first power of x, and considering it as having the exponent 1, or as being x^1, this rule includes the case where x itself is one of the factors. Thus,

$$x^7 \times x = x^8 \text{ or } x^7 \times x^1 = x^{7+1} = x^8$$

Examples. $x^4 \times x^{10} = x^{14}$ $\qquad x^2 \times x^{15} = x^{17}$

$$x \times x^2 \times x^3 \times x^4 = x^{10} \qquad a^6 \times a^6 = a^{12}$$
$$3ab^2 \times 4a^2b = 12a^3b^3 \qquad 2ab \times ab = 2a^2b^2$$
$$a^2b^3ce^6f^3 \times ab^4ce^7f^3 = a^3b^7c^2e^{13}f^6$$

the latter term, written at full length, would be

$$aaabbbbbbbcceeeeeeeeeeeeeffffff$$

To divide a power by another power of a less exponent, *subtract the exponent of the divisor from that of the dividend.* For instance, what is x^{10} divided by x^3? Since 10 is 7 + 3, or since $x^{10} = x^7 \times x^3$ $(\div) x^3$ $\dfrac{x^{10}}{x^3} = x^7$ or x^{10-3}. Similarly, $x^3 \div x$ (or x^1) $= x^2$; $x^{12} \div x^{11} = x^1$ or x; $x^{14} \div x^9 = x^5$; $x^{a+b} \div x^a = x^b$; $a^2b^2c^3 \div abc = abc^2$.

Anomaly 1. If we apply this rule to the division of x^a by x^b, we have $x^a \div x^b = x^{a-b}$. If it should be afterwards found that $a = b$, the preceding result becomes x^{a-a} or x^0, a symbol which as yet has no meaning. We return, therefore, to the original operation, which is, to divide x^a by x^b where $b = a$, or, which is the same thing, to divide x^a by x^a. The answer is evidently 1.

Now why, instead of 1, the rational answer, did we obtain x^0, which has no meaning? Because we applied the preceding rule to a case which does not fall under it. That rule was, "to divide a power by another power of a *less* exponent," &c., and was derived from the preceding rule of multiplication, which rule of multiplication did not apply to any cases except where both factors were powers of x, and, consequently, where the exponent of the product was greater than the exponent of either factor; that is, where x^a is the product, and x^b one of the factors, that rule does not apply unless b be less than a.

When we come to this symbol (x^0) we must do one of two things, either, 1. Consider x^0 as shewing that a rule has been used in a case to which it does not apply, strike it out, and write 1 in its place; or, 2. Let x^0 (which as yet has no meaning) stand for 1; in which case the rule does apply, and gives the true result. Therefore, we lay down the following definition.

By any letter with the exponent 0, *such as* a^0, *we mean* 1; *or every quantity raised to the power whose exponent is* 0, *is* 1.

Anomaly 2. If we apply the equation $x^a \div x^b = x^{a-b}$ to a case in which b is greater than a, say $b = a + 6$, the mere rule gives

$$x^a \div x^{a+6} = x^{a-(a+6)} = x^{-6}$$

a result which has no meaning. The reason is as before; a rule has been applied to a case to which it was not meant to apply. To find the rational result, remember that $x^{a+6} = x^a \times x^6$; and

$$\frac{x^a}{x^{a+6}} = \frac{x^a}{x^a \times x^6} = \frac{1}{x^6}$$

the last result being obtained by dividing both numerator and denominator of the preceding fraction by x^a. The following are similar instances; in the first column is the rational process, in the second the (yet) improper extension of the rule:

$\frac{x^3}{x^4} = \frac{x^3}{x^3 \cdot x} = \frac{1}{x}$	$\frac{x^3}{x^4} = x^{3-4} = x^{-1}$
$\frac{x^2}{x^8} = \frac{x^2}{x^2 \cdot x^6} = \frac{1}{x^6}$	$\frac{x^2}{x^8} = x^{2-8} = x^{-6}$
$\frac{x^{17}}{x^{20}} = \frac{x^{17}}{x^{17} \cdot x^3} = \frac{1}{x^3}$	$\frac{x^{17}}{x^{20}} = x^{17-20} = x^{-3}$

To make the rule applicable, we must agree that

$$x^{-1}, \quad x^{-6}, \quad x^{-3}$$

(which as yet have no meaning) shall stand for

$$\frac{1}{x}, \quad \frac{1}{x^6}, \quad \text{and} \quad \frac{1}{x^3}$$

that is, instead of x^{-1} being the sign that the result is to be abandoned, and $1 \div x$ written in its place, we agree that x^{-1} shall mean the same as $1 \div x$. And we are at liberty to give x^{-1} any meaning we please, because as yet it is without meaning. We lay down, therefore, the following definition:

A letter with a negative exponent means unity divided by the same letter with the same numerical *exponent taken positively; or,*

$$x^{-a} \quad \textit{means} \quad \frac{1}{x^a}$$

Our two rules for multiplication and division will now be found to be universal. The following are instances, arranged as before:

$\frac{1}{x^3} \div \frac{1}{x^8} = \frac{1}{x^3} \times \frac{x^8}{1} = \frac{x^8}{x^3} = x^5$	$x^{-3} \div x^{-8} = x^{-3-(-8)} = x^{-3+8} = x^5$
$x^4 \times \frac{1}{x^6} = \frac{x^4}{x^6} = \frac{1}{x^2}$	$x^4 \times x^{-6} = x^{4+(-6)} = x^{4-6} = x^{-2}$
$\frac{1}{x} \div x^4 = \frac{1}{x^3} \times \frac{1}{x^4} = \frac{1}{x^{3+4}} = \frac{1}{x^7}$	$x^{-3} \div x^4 = x^{-3-4} = x^{-7}$

We have now added to our first definition of an exponent in such a way that two fundamental rules remain true in all cases where the exponents are any whole numbers, positive or negative. These rules are,

$$x^a \times x^b = x^{a+b} \qquad\qquad x^a \div x^b = x^{a-b}$$

But we have no meaning for a number or letter with a *fractional* exponent, such as

$$x^{\frac{1}{2}} \quad x^{\frac{1}{3}} \quad x^{\frac{2}{3}} \quad x^{2\frac{1}{2}} \quad x^{\cdot 61} \quad \text{\&c.}$$

Instead of waiting until some improper extension of the preceding rules shall force us to the consideration of the manner in which it will be most convenient to give meaning to the preceding symbols, we shall endeavour to anticipate this step. And first we shall ask what should $x^{\frac{1}{2}}$ mean, in order that the preceding rule may apply to it? In this case, since $\frac{1}{2}+\frac{1}{2}=1$, we must so interpret $x^{\frac{1}{2}}$ that

$$x^{\frac{1}{2}} \times x^{\frac{1}{2}} = x^{\frac{1}{2}+\frac{1}{2}} = x^1 \quad \text{or} \quad x$$

consequently, $x^{\frac{1}{2}}$ is that quantity which, multiplied by itself, gives x, or is what is called in arithmetic the *square root* of x. Similarly, since $\frac{1}{3}+\frac{1}{3}+\frac{1}{3}=1$, we must so interpret $x^{\frac{1}{3}}$ that

$$x^{\frac{1}{3}} \times x^{\frac{1}{3}} \times x^{\frac{1}{3}} = x^{\frac{1}{3}+\frac{1}{3}+\frac{1}{3}} = x^1 \text{ or } x$$

that is, $x^{\frac{1}{3}}$ must be the cube root of x. By a *root* of x we mean the inverse term to *power;* that is, if m is the *third* power of n, n is called the third root of m. As follows:

Name of n.	Equation implied in the foregoing name.
Square root of m	$nn = m$
Cube root of m	$nnn = m$
Fourth root of m	$nnnn = m$
Fifth root of m	$nnnnn = m$
&c.	&c.

Thus, 4096 is a number which has exact square, cube, fourth, and sixth, and twelfth roots.

Its square root is 64	because	$64 \times 64 = 4096$
Its cube root is 16	....	$16 \times 16 \times 16 = 4096$
Its fourth root is 8	....	$8 \times 8 \times 8 \times 8 = 4096$
Its sixth root is 4	....	$4 \times 4 \times 4 \times 4 \times 4 \times 4 = 4096$
Its twelfth root is 2	....	$\left\{\begin{matrix} 2\times2\times2\times2\times2\times2 \\ \times2\times2\times2\times2\times2\times2 \end{matrix}\right\} = 4096$

These results should, if the preceding interpretation can be relied on be thus expressed:

$$64 = (4096)^{\frac{1}{2}} \qquad 8 = (4096)^{\frac{1}{4}}$$

$$16 = (4096)^{\frac{1}{3}} \qquad 4 = (4096)^{\frac{1}{6}}$$

$$2 = (4096)^{\frac{1}{12}}$$

But, according to the notation best known in arithmetic, they would* be thus expressed:

* $\surd$ is derived from the letter r, the initial of *radix*, or *root*. This symbol is now generally used for the square root, which, in ninety-nine out of a hundred of the applications of algebra, is the highest root which will occur.

$$64 = \sqrt{4096} \qquad 8 = \sqrt[4]{4096}$$
$$16 = \sqrt[3]{4096} \qquad 4 = \sqrt[6]{4096}$$
$$2 = \sqrt[12]{4096}$$

Proceeding with the interpretation of the fractional exponents, $x^{\frac{2}{3}}$ ought to signify the cube root of x^2; for, if the preceding rules are to remain true, we must have

$$x^{\frac{2}{3}} \times x^{\frac{2}{3}} \times x^{\frac{2}{3}} = x^{\frac{2}{3}+\frac{2}{3}+\frac{2}{3}} = x^2$$

and, by the same sort of reasoning, we may conclude that $x^{\frac{m}{n}}$ should stand for the nth root of x^m. But here we may shew that we cannot decide upon the propriety of the preceding interpretation without some further acquaintance with the connexion between roots and powers. For instance,

$$x^{2\frac{1}{2}} \text{ or } x^{2+\frac{1}{2}} \text{ ought to be } x^2 \times x^{\frac{1}{2}} \text{ or } x^2\sqrt{x}$$

But $2\frac{1}{2}$ is $\frac{5}{2}$; therefore,

$$x^{2\frac{1}{2}} \text{ ought to be } x^{\frac{5}{2}} \text{ or } \sqrt{x^5}$$

Consequently, $\qquad x^2 \sqrt{x}$ ought to be $\sqrt{x^5}$

where, by "ought to be," we mean that if it be not so, we cannot,* under the preceding interpretation, apply the common rules of arithmetic to our assumed fractional exponents. Again, since $\frac{1}{2}+\frac{1}{3}=\frac{5}{6}$,

$$x^{\frac{1}{2}} \times x^{\frac{1}{3}} \text{ or } \sqrt{x} \times \sqrt[3]{x} \text{ ought to be } x^{\frac{5}{6}} \text{ or } \sqrt[6]{x^5}$$

but as yet we have neither proved that

$$x^2 \sqrt{x} = \sqrt{x^5} \text{ or that } \sqrt{x} \times \sqrt[3]{x} = \sqrt[6]{x^5}$$

For the purpose of shewing that the conjectural interpretation as yet given leads to no erroneous results, we premise the following *arithmetical* theorems.

Theorem I. If a be *greater* than b, than a^2 is *greater* than b^2, a^3 than b^3, &c. For aa is then the result of a multiplication in which more than b is taken more times than there are units in b; therefore aa is greater than bb: in a^3 or a^2a, more than b^2 is taken more times

* The student must remember that we are perfectly free to make $x^{2\frac{1}{2}}$ and $x^{\frac{5}{2}}$ mean different things, if we only take care not to confound the two. But it would be inconvenient that $2\frac{1}{2}$ should any where have a meaning different from that of $\frac{5}{2}$.

than there are units in b, and so on. By changing the order in which the letters are named, the theorem may be differently worded, thus: if b be less than a, then b^2 is less than a^2, &c.

Theorem II. If a be *greater* than b, a^{-1} is *less* than b^{-1}, a^{-2} is *less* than b^{-2}. For if a be greater than b, $\frac{1}{a}$ is less than $\frac{1}{b}$; and since, in that case, a^2 is greater than b^2, therefore $\frac{1}{a^2}$ is less than $\frac{1}{b^2}$, and so on. Similarly, if a be *less* than b, a^{-1} is greater than b^{-1}, &c.

Theorem III. If a be *equal* to b, then a^2 is *equal* to b^2, a^3 to b^3, and so on. This is evident from page 3.

Theorem IV. If a be *equal* to b, the square root of a is *equal* to the square root of b, the cube root of a to the cube root of b, and so on. Let m and n be, for example, the fifth roots of a and b (which last are equal), then a and b are the fifth powers of m and n; if m were the greater of the two, its fifth power a (Theorem I.) would be greater than b, which is not the case. Similarly, if n were the greater of the two, b would be greater than a; therefore m must be equal to n. In the same way any other case may be proved.

Theorem V. If a be *greater* than b, the square root of a is *greater* than the square root of b, &c. (We put the preceding argument* in different words). Since a is the square of its square root, and b the same; if the square root of a were *equal* to the square root of b (by Theorem III.), the square of the first (or a) would be *equal* to the square of the second (or b), which is not the case. If the square root of a were *less* than the square root of b (by Theorem I.), the square of the first (or a) would be *less* than the square root of the second (or b); which is not the case. The only remaining possibility is, that when a is *greater* than b, then the square root of a is *greater* than the square root of b. Similarly, if a be *less* than b, the square root of a is *less* than the square root of b: and so on.

Theorem VI. An arithmetical quantity has but one arithmetical square, cube, or any other root. For suppose a, if possible, to have two *different* cube roots, m and n; one of these two is the greater, let it be m. Then (Theorem I.) the cube of m (or a) is greater than the cube of n (also a); but the last two assertions are contradictory, therefore a cannot have two different cube roots, &c.

* The nature of the argument is supposed to be well understood from the last; it is practice in the use of terms which is here given.

All whole numbers have not whole square roots, or cube roots; and the higher the order of the root, the fewer are the whole numbers lying under any given limit which have a root of the kind. The following table will illustrate this.

Numbers which have a whole					
Square root.	Cube Root.	Fourth Root.	Fifth Root.	Sixth Root.	Value of the Root.
1	1	1	1	1	1
4	8	16	32	64	2
9	27	81	243	729	3
16	64	256	1024	4096	4
25	125	625	3125	15625	5
36	216	1296	7776	46656	6
49	343	2401	16807	117649	7
64	512	4096	32768	262144	8
81	729	6561	59049	531441	9
100	1000	10000	100000	1000000	10
&c.	&c.	&c.	&c.	&c.	&c.

A number which has not a whole root has not an exact fractional root. At present we shall only enunciate the following proposition, without proving it, leaving the student to try if he can produce any instance to the contrary.

No power or root of a fraction can be a whole number.*

Consequently, all those problems of arithmetic or algebra are misconceptions, which require the extraction of any root of a number, unless that number be one of those specified in the preceding table (continued *ad infinitum*) as having such a root. But though we may not look for the *exact* solution of such problems, we shall shew that solutions may be found which are as nearly true answers as we please; that is, we shall prove the following theorem.

Though there is no fraction whose n*th power is exactly any given whole number, we may assign a fraction whose* n*th power shall differ*

* That is, of a real fraction; $\frac{4}{2}$, $\frac{64}{4}$, &c. are whole numbers in a fractional form. It is not necessary to prove this strictly here; because, were it not true, what follows would not be incorrect, but only useless.

from that whole number by less than any quantity named, say ·0001, or ·0000001, *or any other small fraction.*

N.B. With the most convenient method of finding such a fraction we have here nothing to do, but only with the proof that it can be found. This is shewn in arithmetic as to the square and cube root only at most. We must enter into the proof of this at some length, and shall lay down the following *Lemmas.**

Lemma 1. The powers of 2 are formed by one addition: thus,

$$2+2=2^2 \qquad 2^2+2^2=2^3 \qquad 2^3+2^3=2^4$$

or generally, $$2^n+2^n=2^n\times 2=2^{n+1}$$

Lemma 2. The powers of a fraction are formed by forming the powers of the numerator and denominator: thus,

$$\left(\frac{a}{b}\right)^3=\frac{a}{b}\times\frac{a}{b}\times\frac{a}{b}=\frac{aaa}{bbb}=\frac{a^3}{b^3}$$

Lemma 3. If p be less than q
Then ap is less than aq

If p be less than q
and a .... b
ap is less than bq

Lemma 4. If v be less than unity its powers decrease continually. For instance, if v be one half, its square ($\frac{1}{2}\times\frac{1}{2}$) is one half of one half, which is less than v; its cube is one half of one fourth, which is less than v^2; and so on.

Lemma 5. If one of the positive terms of an expression be increased the expression itself is increased, &c. Thus $a-b$ is increased by increasing a, and decreased by decreasing a; but it is *decreased* by *increasing* b, and *increased* by *decreasing* b.

Lemma 6. *If v be less than* 1, then

$(1+v)^2$ is less than $1+3v$ or $1+(4-1)v$
$(1+v)^3$ ...... $1+7v$ or $1+(8-1)v$
$(1+v)^4$ ...... $1+15v$ or $1+(16-1)v$
....
or $(1+v)^n$ is less than $1+(2^n-1)v$

Firstly, $(1+v)^2$ or $(1+v)(1+v)$ is $1+2v+v^2$, which, since (Lemma 4) v is greater than v^2, is (Lemma 5) increased by writing

* A Lemma is a proposition which is only used as subservient to the proof of another proposition.

v for v^2. But it then becomes $1+2v+v$ or $1+3v$. Therefore $1+2v+v^2$ is less than $1+3v$; that is, $(1+v)^2$ is less than $1+3v$.

Again,	$(1+v)^2$	is less than	$1+3v$
(Lemma 3)	$(1+v)^2(1+v)$	......	$(1+3v)(1+v)$
or	$(1+v)^3$	......	$1+4v+3v^2$

Still more, then, (Lemmas 4 and 5 as before) is $(1+v)^3$ less than $1+4v+3v$, or $1+7v$.

Again,	$(1+v)^3$	is less than	$1+7v$
Therefore	$(1+v)^4$	......	$(1+7v)(1+v)$
		or	$1+8v+7v^2$
Still more is	$(1+v)^4$	less than	$1+8v+7v$
		or	$1+15v$

We might thus proceed through any number of steps, but the following is a species of proof which embraces all. Suppose that one of the preceding is true, say that containing the nth power; that is, let

$$(1+v)^n \quad \text{be less than} \quad 1+(2^n-1)v$$

Then (Lem. 3) $(1+v)^n(1+v)$ is less than $\{1+(2^n-1)v\}(1+v)$

which product may be found as follows:

$$\begin{array}{lll} & 1+(2^n-1)v & \\ & 1+\qquad\quad v & \\ \hline & 1+(2^n-1)v & \\ & \qquad\qquad v & +(2^n-1)v^2 \\ \hline \text{Add} & 1+2^n v & +(2^n-1)v^2 \end{array}$$

or $\quad (1+v)^{n+1}$ is less than $1+2^n v+(2^n-1)v^2$

Still more (Lemmas 4 and 5) is it less than . . . . . . $1+2^n v+(2^n-1)v$

or $\quad 1+(2^n+2^n-1)v$

or (Lemma 1) $\quad 1+(2^{n+1}-1)v$

We have proved, then, that

if $\quad (1+v)^n$ be less than $1+(2^n-1)v$

it follows that $(1+v)^{n+1}$ is less than $1+(2^{n+1}-1)v$

or in the series of propositions contained in this lemma, each one must be true if the preceding be true. But the first has been proved, therefore all have been proved.

Lemma 7. *If x be greater than a*, then

$$(x+a)^2 \text{ is less than } x^2+3ax \quad \text{or} \quad x^2+(4-1)ax$$
$$(x+a)^3 \text{ .... } x^3+7ax^2 \quad \text{or} \quad x^3+(8-1)ax^2$$
$$(x+a)^4 \text{ .... } x^4+15ax^3 \quad \text{or} \quad x^4+(16-1)ax^3$$

....

or $(x+a)^n$ is less than $x^n+(2^n-1)ax^{n-1}$

Because x is greater than a, $\frac{a}{x}$ is less than 1; therefore, (Lemma 6),

$$\left(1+\frac{a}{x}\right)^n \text{ is less than } 1+(2^n-1)\frac{a}{x}$$

But $1+\frac{a}{x}=\frac{x+a}{x}$ $\therefore$ (Lemma 2) $\left(1+\frac{a}{x}\right)^n=\frac{(x+a)^n}{x^n}$

Therefore, $\frac{(x+a)^n}{x^n}$ is less than $1+(2^n-1)\frac{a}{x}$

Multiply both sides by x^n (Lemma 3), which gives

$$(+a)^n \text{ is less than } x^n+(2^n-1)\frac{ax^n}{x}$$
$$\text{or } x^n+(2^n-1)ax^{n-1}$$

We proceed to shew the proposition in pages 90 and 91 in a particular case. Say the number is 10, the power mentioned is the cube. Can a fraction be found whose cube shall be within, say ·0001 of 10? Since $(2)^3=8$, and $(3)^3=27$, 2 is too small and 3 too great. Examine the cubes of the following fractions falling between 2 and 3; 2·1, 2·2, 2·3, &c. We have $(2{\cdot}1)^3=9{\cdot}261$, and $(2{\cdot}2)^3=10{\cdot}648$; whence 2·1 is too small, and 2·2 too great. Examine the fractions 2·11, 2·12, 2·13, &c. lying between 2·1 and 2·2. We find,

$$(2{\cdot}15)^3=9{\cdot}938375 \qquad (2{\cdot}16)^3=10{\cdot}077696$$

Therefore 2·15 is too small, and 2·16 too great.

Proceeding in this way, we shall find,

$(2{\cdot}154)^3$	less than	10	$(2{\cdot}155)^3$	greater than	10
$(2{\cdot}1544)^3$	..	10	$(2{\cdot}1545)^3$	..	10
$(2{\cdot}15443)^3$	..	10	$(2{\cdot}15444)^3$	..	10
&c.			&c.		

The only question now is, shall we thus arrive at two fractions, one having a cube less than 10, and the other greater, but both cubes so near to 10 as not to differ from it by ·0001? Observe that in the preceding list,

2·2	exceeds	2·1	by only	·1
2·16	..	2·15	..	·01
2·155	..	2·154	..	·001
2·1545	..	2·1544	..	·0001
&c.		&c.		&c.

and from Lemma 7, if a be less than x.

$$(x+a)^3 \text{ is less than } x^3+7ax^2$$

$$\text{or } (x+a)^3-x^3 \text{ is less than } 7ax^2$$

Let x be the lower of one of the preceding sets of fractions; then, since x^3 is less than 10, x must be less than 3, and its square less than 9. Therefore, $7ax^2$ must be less than $7a\times 9$, or than $63a$. Still more, then, will $(x+a)^3-x^3$ (which is less than $7ax^2$), be less than $63a$. Let $x+a$ be the higher of the fractions in the set spoken of; then a, the difference, will at some one step become ·0000001, consequently, $63a$ will become ·0000063, which is less than ·00001. Therefore x may be so found that

$$x^3 \text{ is less than } 10 \qquad (x+\cdot0000001)^3 \text{ is greater than } 10$$

$$\text{and } (x+\cdot0000001)^3-x^3 \text{ is less than } \cdot00001.$$

But 10, which lies between the two cubes, will differ from either of the cubes by less than they differ from each other; therefore, either fraction has a cube within the required degree of nearness to 10. The fractions required would be found to be 2·1544346 and 2·1544347. In a similar way any other case might be treated.

Hence, the following language is used. Instead of saying that 10 has no cube root, but that fractions may be found having cubes as near to 10 as we please, those fractions are called *approximations** to the cube root of 10, as if there were such a thing as $\sqrt[3]{10}$. Thus, 2·154 is an approximation to $\sqrt[3]{10}$, but not so near an approximation as 2·1544346; instead of saying that $(2{\cdot}154)^3$ is nearly equal to 10, but not so nearly equal to 10 as $(2{\cdot}1544346)^3$.

* *Approximare*, to bring near to.

The student will now understand the sense in which we use the following words:

Every number and fraction has a root of every order, either exact or approximate.

When we shall have proved that in all cases $\sqrt{a} \times \sqrt[3]{a} = \sqrt[6]{a^5}$, what do we mean by this equation in the case where $a = 10$, and therefore $a^5 = 100000$? We mean that we can obtain two fractions of which the square and cube are within any degree of nearness (say ·0001) of 10, which we call approximate values of $\sqrt{10}$ and $\sqrt[3]{10}$, and that, on multiplying these two fractions together, we find a product which, being raised to the sixth power, gives a result within the same degree of nearness to 10^5, or is an approximate value of $\sqrt[6]{10^5}$. We shall anticipate the proof of both propositions, as one specimen of the method of passing from the strict to the approximative proposition will serve for all.

Let a be a number which has both a square and a cube root (such as 64, or $\frac{1}{729}$). Let x be the square root, and y the cube root. Then

$$x^2 = a \text{ therefore } (x^2)^3 = a^3 \text{ or } x^2.x^2.x^2 = a^3 \text{ or } x^6 = a^3$$
$$y^3 = a \text{ therefore } (y^3)^2 = a^2 \text{ or } y^3.y^3 = a^2 \text{ or } y^6 = a^2$$
$$\therefore \quad x^6y^6 = a^3a^2 \text{ or } (xy)^6 = a^5$$

for x^6y^6 is $xxxxxxyyyyyy$, in which the multiplications may be performed in the order

$$xy.xy.xy.xy.xy.xy \text{ giving } (xy)^6$$

Consequently, xy is the sixth root of a^5; but x is the square root of a, and y the cube root, that is,

$$xy = \sqrt[6]{a^5} \quad \text{or} \quad \sqrt{a} \times \sqrt[3]{a} = \sqrt[6]{a^5}$$

Now, suppose a to be a number which has neither square or cube root, such as 10. We can find fractions x and y, such that x^2 and y^3 shall be as near as we please to a. Say that $x^2 = a + p$ and $y^3 = a + q$ where p and q may be as small* as we please. We will begin by supposing p and q smaller than a. Hence (Lemma 7),

* *As small as we please* does not mean that we can choose them exactly, but that, name any fraction we may, however small, they may be made (we need not inquire how much) smaller.

$$(a+p)^3 \text{ is less than } a^3+7pa^2$$
$$(a+q)^2 \quad \cdots\cdots \quad a^2+3qa$$

But $(x^2)^3$ or $x^6 = (a+p)^3$ $\quad (y^3)^2$ or $y^6 = (a+q)^2$
therefore,

$$x^6 \text{ is less than } a^3+7pa^2$$
$$y^6 \quad \cdots\cdots \quad a^2+3qa$$

(Lemma 3) $x^6y^6 \quad \cdots\cdots \quad (a^3+7pa^2)(a^2+3qa)$

or $\quad (xy)^6$ is less than $a^5+(7p+3q)a^4+21pqa^3$

But since x^2 (being $a+p$) is greater than a, x^6 is greater than a^3; and since y^3 (being $a+q$) is greater than a, y^6 is greater than a^2; hence x^6y^6 or $(xy)^6$ is greater than $a^3.a^2$ or a^5. Hence,

$$(xy)^6 \text{ lies between } a^5 \text{ and } a^5+(7p+3q)a^4+21pqa^3$$

and therefore does not differ from a^5 by so much as

$$(7p+3q)a^4+21pqa^3$$

Now, since p and q may be as small as we please, $7p+3q$ and $21pq$ may be as small as we please, and, therefore (however great* a^4 and a^3 may be), may be so taken that the preceding expression shall be as small as we please. That is, $(xy)^6$ may be made as near an approximation as we please to a^5, or xy is an approximate sixth root of a^5.

The preceding demonstration is not frequently given in books on algebra, but the result is assumed in what is called the *law of continuity*. This term we shall proceed to explain, as it involves considerations which will be useful to the student; but as it may be omitted without breaking the series of results, we inclose it in brackets, and also the heading of the pages which contain it.

[The word *continuous* is synonymous with *gradual* or *without sudden changes*. For instance, suppose a large square, with two of its opposite sides running north and south. A person who walks round this square will make a quarter-face at each corner, that is, will at once proceed east or west where before he was moving north or south, and *vice versâ*, without moving in any of the intermediate directions.

* The product mn, if n be given, and m as small as we please, may be made as small as we please; only the greater n is, the smaller must m be taken, in order to give the product the required degree of smallness.

In this case he changes his direction *discontinuously*. If the figure had had eight sides he would still have made discontinuous changes of direction, but each change of less amount; still less would the changes have been if the figure had had sixteen sides, and so on.

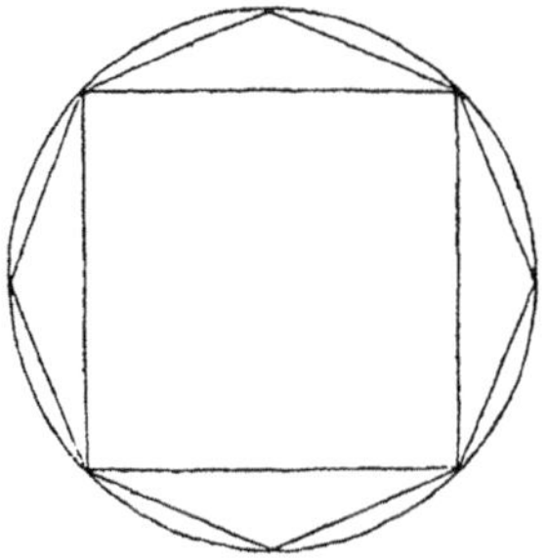

But if he walk round a circle, or other oval curve, there is no discontinuous change of direction. If a geometrical* point move round a geometrical circle, there is no conceivable direction in which it will not be moving at some point or other of its course; and between every two points, it will move in every direction intermediate to the directions which it has at those two points.

The preceding illustration is drawn from geometry, in which continuous change is supposed; and there is nothing repugnant to our ideas in the supposition, but the contrary, at least, when we imagine lines to be created by the motion of a point. But in the application of arithmetic, and eventually of algebra, to geometry, we have this question to ask, Can all geometrical magnitudes be represented arithmetically? For instance, suppose B to start from A, and to move in a straight line till it is 100 feet from A, can we, by means of feet and fractions of feet, represent the distance to which B has moved from A, in every one of the infinite number of points which B passes through?

It is clear that between one foot and two feet we can interpose the fractions 1·1, 1·2, 1·3, &c. feet; between 1·1 and 1·2 feet we can interpose 1·11, 1·12, 1·13, &c. feet; between 1·11 and 1·12 we can

* *Geometrical*, formed with the accuracy which the reason supposes in geometrical figures.

interpose 1·111, 1·112, 1·113, &c. feet; and so on for ever.* But assigning B any geometrical position between 1 and 2 feet distance from A, we cannot be prepared to say that we shall thus come at last to a foot and a fraction of a foot, which will exactly represent that position. And we shall now proceed to shew one position, at least, assignable geometrically, but not arithmetically.

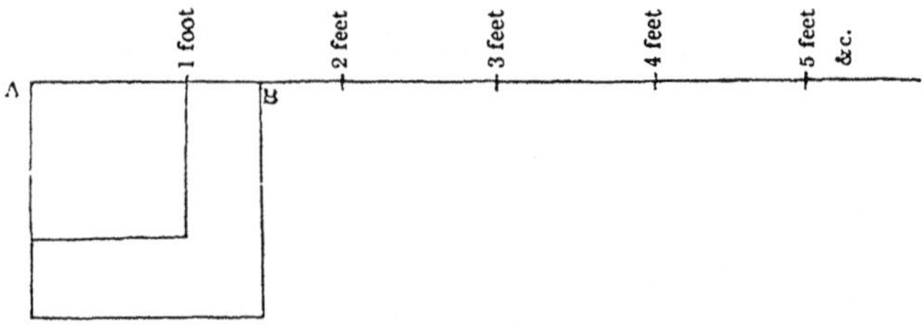

It is shewn in geometry† how to assign (by geometrical construction, not arithmetically) a position to B, in which the square described on AB shall be twice as great as the square described on 1 foot (as in the figure). We now proceed to inquire whether the line AB has, in such a case, any assignable arithmetical magnitude (a foot being represented by 1). If so, since every fraction of a foot can be reduced to a fraction with a whole numerator and denominator (Ar. 114, 121), let AB be $\frac{m}{n}$ feet, where m and n are whole numbers. That is, let AB be formed by dividing one or more feet each into n equal parts, and putting together m of those parts. Let this nth part of a foot be called for convenience a "subdivision;" then a foot contains n subdivisions, and AB contains m subdivisions. Then, from Ar. 234, it appears that the square on the foot contains $n \times n$ of the squares described on a subdivision, and the square on AB contains $m \times m$ of the same. Hence, since the square on AB is double of the square on one foot, we must have

$$mm = 2nn$$

We now proceed to shew that this equation is not possible under the stipulation that m and n are whole numbers. Because n is a

* The student must not imagine this phrase, or its corresponding Latin *ad infinitum*, to mean more than as long as we please, or as far as we please.

† Euclid, book ii. last proposition.

whole number, nn is a whole number, and $2nn$ is twice a whole number, and therefore an even number; but mm equals twice nn, therefore mm is even. Therefore m is even, for an odd number multiplied by itself gives an odd number. But if m be even, its half is a whole number; let that half be m', then $m = 2m'$. Substitute this value of m in the equation, which gives

$$2m' \times 2m' = 2nn \qquad 4m'm' = 2nn \quad \text{or} \quad 2m'm' = nn$$

which last equation $nn = 2m'm'$ may be used in a manner precisely similar, to shew that n must be an even number. Let its half be n' (a whole number), then $n = 2n'$, and substitution gives

$$2n' \times 2n' = 2m'm' \qquad 4n'n' = 2m'm' \quad \text{or} \quad 2n'n' = m'm'$$

which proves as before that m' is even. In this way we shew that in order that the equation $mm = 2nn$ may be true (m and n being whole numbers), we must have

$$m \quad (m' \text{ or half of } m) \quad (m'' \text{ or half of } m') \text{ \&c.}$$
$$n \quad (n' \text{ or half of } n) \quad (n'' \text{ or half of } n') \text{ \&c.}$$

all even whole numbers for ever. But this cannot be; for if any number be halved, if its half be halved, and so on, we shall at last come to a fraction less than 1. Consequently, the equation $mm = 2nn$ cannot be true of any whole numbers, and therefore AB cannot be represented by any fraction $\frac{m}{n}$.

The preceding equation (if it could exist) would give

$$\frac{mm}{nn} = 2 \quad \text{or} \quad \frac{m}{n} \times \frac{m}{n} = 2 \quad \text{or} \quad xx = 2 \quad \text{where} \quad x = \frac{m}{n}$$

and we have seen (page 94) that we can admit the equation $xx = 2$ only in this sense, that, naming any fraction, however small, we can find a value for x, which shall give xx differing from 2 by less than that fraction. That is, instead of satisfying the equation

$$xx - 2 = 0$$

we can only satisfy the equation

$$xx - 2 = \text{a quantity arithmetically less than} \ (\qquad)$$

where we may fill up the blank with any fraction we please, however small.

This is sufficient for all practical purposes; because no application of algebra which has any reference to the purposes of life can require a degree of accuracy beyond the limits of our sight, when assisted by the most accurate means of measurement. If we take the least possible visible line to be that of a ten thousandth of an inch, it will certainly be sufficiently near the truth to solve the equation

$xx-2 =$ arithmetically less than one millionth of an inch,
in cases where perfect exactness demands that $xx-2 = 0$

The solution which nearly satisfies such an equation as $xx-2=0$ may be either too small or too great; that is, xx may be a little less or a little greater than 2. See page 94, where sets of solutions of both kinds are given for the equation $xxx-10=0$. Hence, though x or $\sqrt[3]{10}$ has no existence, yet, since we can find two fractions, say a and b, as near to one another as we please, of which the first is too small (or aaa less than 10) and the second too great (or bbb greater than 10); and since we can carry this process to any degree of accuracy short of positive exactness, it is usual to make use of such forms of speech as the following: 10 has a cube root, but that cube root is an *incommensurable* * quantity, not expressible by any number or fraction, except approximately; that is, we can find a fraction as near as we please to $\sqrt[3]{10}$.

But still the following question remains: though we can, *quam proximè*,† solve the equation $xx-2=0$, may there not be processes to which it may be necessary to subject that approximate solution, and may not those processes have this property, that any error, however small, in the quantity to which they are applied, creates an error which cannot be diminished beyond a certain extent, however small the original error may be? For instance, when the student comes to know what is meant by the *logarithm* of a number (for our present purposes it suffices to say, that it is the result of a complicated process), he might happen to meet with a problem the answer to which is "the logarithm of x, where x is to be found from the equation $xx-2=0$." Which of the two following propositions will he take? one of them must be true.

1. "In taking the logarithm of x, the error committed in finding

* Having no common measure with 1; not expressible by adding together any of the subdivisions of 1.

† A Latin phrase frequently applied to this subject, and therefore introduced here. Translate it, *as near as we please*.

x may be made so small, that the error in the logarithm shall be less than any fraction named."

2. "If an error committed in finding a number be ever so small, the error of the logarithm must be greater than (some given fraction, say) ·001."

To this no answer can be given except the following caution :

Whenever any new process is introduced, or any new expression, it must be proved, and not assumed, that problems involving that process admit of quam proximè, *if not of exact, solutions.*

The student should now apply himself to prove this of the processes already described. We will take one case at length.

Let the result of a problem be $\dfrac{a+b}{c+e}$

where an error, which we are at liberty to suppose as small as we please, has been committed in determining, say b and e. Or (which is a way of speaking more consistent with what has preceded), let b and e be involved in some equations (such as $bb-2=0$, $eee-3=0$) which only admit *quam proximè* solutions. Let b' and b'' be approximate values of b, the first too small, the second too great; let e' and e'' be approximate values of e. Then the substitution of the approximate values of the preceding expression gives

$$\frac{a+b'}{c+e'}, \quad \text{and} \quad \frac{a+b''}{c+e''}$$

To compare these expressions, subtract the second from the first, which gives

$$\frac{a+b'}{c+e'}-\frac{a+b''}{c+e''}=\frac{(ac+ae''+cb'+b'e'')-(ac+ae'+cb''+e'b'')}{cc+ce'+ce''+e'e''}$$

$$=\frac{a(e''-e')-c(b''-b')+(b'e''-e'b'')}{cc+(e'+e'')c+e'e''}$$

And since, page 94, e'' may be brought as near to e' as we please, and b'' to b', it follows that $e''-e'$ and $b''-b'$ may be made as small as we please; whence $a(e''-e')$ and $c(b''-b')$ may be made as small as we please, page 96, note. And so may $b'e''-e'b''$, for it will be found to be the same as

$$b'(e''-e')-e'(b''-b')$$

Hence the numerator of the preceding fraction can be made as small as we please, because each of its terms can be made as small as we please. But since e'' is always greater than e', the denominator is (page 91, Lemma 5) always greater than

$$cc + (e' + e')c + e'e'$$

Here then is a fraction of which the numerator can be made as small as we please, but not the denominator; consequently, the fraction can be made as small as we please. That is, the fractions

$$\frac{a+b'}{c+e'} \text{ and } \frac{a+b''}{c+e''}, \text{ different values attributed to } \frac{a+b}{c+e}$$

can be brought as near as we please, or be made to differ as little as we please. Here, by the same extension of language as before, the last mentioned fraction is said to have a definite value, to which we approximate by substituting approximate values of b and e.

The *law of continuity* which is assumed to exist in algebraical expressions, and which must be proved by the student in a sufficient number of particular cases, consists in the following theorem:

GENERAL THEOREM.

Let there be an algebraical expression P, which contains x, and let the substitution of a instead of x give to that expression the value p.

Let the substitution of $a+m$ instead of x give to the expression P the value q.

Then if a and $a+m$ may be made as nearly equal as we please; that is, if m may be made as small as we please, it will always follow that p and q may be made to differ as little as we please.

PARTICULAR CASE.

$$\text{P} = x + x^2$$
$$p = a + a^2$$

$$q = (a+m) + (a+m)^2$$
$$= a + a^2 + (1+2a)m + m^2$$
$$= p \quad + (1+2a)m + m^2$$

From the last,

$$q - p = (1+2a)m + m^2$$

each term of which, if m may be made as small as we please, may be made as small as we please.

The only question about which any doubt can arise, as regards expressions hitherto obtained, relates to the values, real or approximate, as the case may be, of $\sqrt{x}$, $\sqrt[3]{x}$, &c. If m can be made as small as we please, can the approximate values of $\sqrt[3]{a}$ and $\sqrt[3]{a+m}$ be made as near as we please? To answer this question, we must premise a theorem which will also be useful in many other places.

It is as follows:

$$x^2-y^2 = (x-y)(x+y)$$
$$x^3-y^3 = (x-y)(x^2+xy+y^2)$$
$$x^4-y^4 = (x-y)(x^3+x^2y+xy^2+y^3)$$
$$\dots\dots\dots\dots\dots\dots\dots\dots$$
$$x^n-y^n = (x-y)(x^{n-1}+x^{n-2}y+\cdots+xy^{n-2}+y^{n-1})$$

Multiplication will make any one of these obvious; for instance,

$$\begin{array}{l} x^3+x^2y+xy^2+y^3 \\ x-y \\ \hline x^4+x^3y+x^2y^2+xy^3 \\ \quad -x^3y-x^2y^2-xy^3-y^4 \\ \hline x^4+0+0+0-y^4 \end{array}$$

(Observe that in this, the first multiplication which has been given at length since Chapter II., we have employed the second of the methods in the Introduction, and shall do so in future.)

If we examine the series of expressions

$$x+y, \quad x^2+xy+y^2, \quad x^3+x^2y+xy^2+y^3, \text{ \&c.}$$

we shall see that each of them may be made by multiplying the preceding by y, and adding a new power of x. Thus,

$$x^2+xy+y^2 = x^2+y(x+y)$$
$$x^3+x^2y+xy^2+y^3 = x^3+y(x^2+xy+y^2) \text{ \&c.}$$

So that, if we call* these expressions P_1, P_2, P_3, &c. (we shall seldom use large letters except as the abbreviations of other expressions), we have

$$P_2 = x^2+yP_1, \quad P_3 = x^3+yP_2, \quad P_4 = x^4+yP_3, \text{ \&c.}$$

or generally $\quad P_n = x^n+yP_{n-1}$

But we shall also find that the same expressions may be made by multiplying by x, and adding powers of y, as follows:

$$x^2+xy+y^2 = y^2+x(x+y)$$
$$x^3+x^2y+xy^2+y^3 = y^3+x(x^2+xy+y^2) \text{ \&c.}$$

* The figures underwritten must not be confounded with exponents. They are used as the accents in page 38, and are read *P one*, *P two*, *P three*, &c.

or $P_2 = y^2 + xP_1 \quad P_3 = y^3 + xP_2 \quad P_4 = y^4 + xP_3$ &c.

or generally, $P_n = y^n + xP_{n-1}$

The theorem we are now upon may be thus expressed:

$$x^n - y^n = (x-y)P_{n-1}$$

and it may be proved as follows:

General Theorem.	Particular Case.
$P_n = x^n + yP_{n-1}$	$P_4 = x^4 + yP_3$
$P_n = y^n + xP_{n-1}$	$P_4 = y^4 + xP_3$
$(-)\ 0 = x^n - y^n - (x-y)P_{n-1}$	$(-)\ 0 = x^4 - y^4 - (x-y)P_3$
$x^n - y^n = (x-y)P_{n-1}$	$x^4 - y^4 = (x-y)P_3$

Let us now examine $\sqrt[3]{10}$ and $\sqrt[3]{10+m}$, where m may be made as small as we please. Let y and x be approximate values of the first and second, so that (pages 94, 95),

$$\left.\begin{array}{l} x^3 = (10+m) + v \\ y^3 = 10 \qquad\quad + w \end{array}\right\} \text{ where } v \text{ and } w \text{ may be made as small as we please.}$$

$$x^3 - y^3 = m + v - w$$

or $$(x-y)(x^2 + xy + y^2) = m + v - w$$

$$x - y = \frac{m + v - w}{x^2 + xy + y^2}$$

Now x and y are both greater than 2, since $2^3 = 8$ (less than 10): therefore the denominator of the preceding fraction must be greater than 12, while the numerator can be made as small as we please. Hence the fraction (which is $= x - y$) can be made as small as we please, or x and y as nearly equal as we please. But x and y are the approximate values of $\sqrt[3]{10+m}$ and $\sqrt[3]{10}$.

The student may try to prove the following theorem:

$$x^2 - y^2 = (x+y)(x-y)$$
$$x^4 - y^4 = (x+y)(x^3 - x^2y + xy^2 - y^3)$$
$$x^6 - y^6 = (x+y)(x^5 - x^4y + x^3y^2 - x^2y^3 + xy^4 - y^5)$$

&c. &c.

$$x^3 + y^3 = (x+y)(x^2 - xy + y^2)$$
$$x^5 + y^5 = (x+y)(x^4 - x^3y + x^2y^2 - xy^3 + y^4)$$

&c. &c.

We shall now proceed with the general theory of exponents.]

To raise a power of x *to any power, multiply the exponents of the two powers together for an exponent;* for instance,

$$(x^3)^4 = x^{3\times 4} = x^{12} \text{ for } (x^3)^4 = x^3.x^3.x^3.x^3 = x^{3+3+3+3}$$

In a similar manner,

$$(x^2)^{12} = x^{24} \qquad (x^3)^6 = x^{18} \qquad (x^a)^b = x^{ab}$$

$$(x^{a+b})^{a-b} = x^{a^2-b^2} \qquad (x^{a-b})^{a+b} = x^{a^2-b^2}$$

$$(x^m)^n = (x^n)^m = x^{mn}$$

A power of a product is the product of the powers of the factors. Thus,

$$(abc)^3 = abc.abc.abc = aaabbbccc = a^3b^3c^3$$

$$(ab^2c^3)^4 = a^4(b^2)^4(c^3)^4 = a^4b^8c^{12}$$

$$(a^mb^nc^p)^r = (a^m)^r(b^n)^r(c^p)^r = a^{mr}b^{nr}c^{pr}$$

The same rule may be applied to division. Thus, $\left(\frac{a}{b}\right)^3 = \frac{a^3}{b^3}$; for the first is $\frac{a}{b}\times\frac{a}{b}\times\frac{a}{b}$, or $\frac{aaa}{bbb}$, which is the second (page 91).

A root of a root is that root whose index is the product of the indices of the first mentioned roots. Thus, the fourth root of the cube (third) root is the twelfth root. To prove this, let the fourth root of the third root of x be called y. Then,

$$\sqrt[4]{\sqrt[3]{x}} = y \;\therefore\; \sqrt[3]{x} = y^4 \qquad x = (y^4)^3 = y^{12}$$

$$\text{or} \quad y = \sqrt[12]{x}, \quad \text{but } y \text{ is also } \sqrt[4]{\sqrt[3]{x}}$$

We have shewn, page 89, that x can have but one arithmetical cube root, or twelfth root; and that the cube root can have but one arithmetical fourth root. Hence the above process is conclusive; it shews that *a* fourth root of *a* cube root of x is *a* twelfth root of x; and there is but one arithmetical root of each kind. But when we come to consider all the algebraical symbols which are roots of x, both those which have arithmetical meaning and those which have not, the student must remember that the preceding does not prove that *every* fourth root of *every* third root of x is a twelfth root of x. This may be the case, but it is not yet proved.

In a similar way it may be proved that

$$\sqrt{\sqrt{x}} = \sqrt[4]{x}; \quad \sqrt{\sqrt[3]{x}} = \sqrt[6]{x} = \sqrt[3]{\sqrt{x}}$$

$$\sqrt[a]{\sqrt[b]{x}} = \sqrt[ab]{x} = \sqrt[b]{\sqrt[a]{x}}; \quad \sqrt[5]{\sqrt[5]{x}} = \sqrt[25]{x}$$

The root of a product is found by multiplying together the roots of the factors. Thus,

$$\sqrt[4]{abc} = \sqrt[4]{a} \times \sqrt[4]{b} \times \sqrt[4]{c} \quad \cdots\cdots\cdots\cdots \quad \text{(A)}$$

for these two have the same fourth power, namely, abc. For by definition (page 87),

$$(\sqrt[4]{abc})^4 = abc$$

and by page 105,

$$(\sqrt[4]{a} \times \sqrt[4]{b} \times \sqrt[4]{c})^4 = (\sqrt[4]{a})^4 \times (\sqrt[4]{b})^4 \times (\sqrt[4]{c})^4 = a \times b \times c$$

That is, each side of the equation (A) is a fourth root of abc. But abc has but one arithmetical fourth root; consequently, each side of (A) must be that root; and therefore the two sides are equal. Similarly it may be proved that

$$\sqrt{abc} = \sqrt{a} \times \sqrt{b} \times \sqrt{c} \qquad \sqrt{ab^2} = \sqrt{a} \times \sqrt{b^2} = b\sqrt{a}$$

$$\sqrt[3]{a^2b^2c^4} = \sqrt[3]{a^2} \times \sqrt[3]{b^2} \times \sqrt[3]{c^4} \qquad \sqrt[4]{32} = \sqrt[4]{16} \times \sqrt[4]{2} = 2\sqrt[4]{2}$$

The same rule may be applied to division. Thus, $\sqrt[3]{\frac{a}{b}} = \frac{\sqrt[3]{a}}{\sqrt[3]{b}}$;

for both of these will be found to have the same cube, namely, $\frac{a}{b}$.

If a power of x *be raised, and a root of that power extracted, the result is not altered if the order of the operations be changed.* That is,

$$\sqrt[3]{x^4} \text{ is the same as } \left\{\sqrt[3]{x}\right\}^4$$

To prove this, observe that (page 105),

$$\sqrt[3]{x^4} = \sqrt[3]{xxxx} = \sqrt[3]{x} \times \sqrt[3]{x} \times \sqrt[3]{x} \times \sqrt[3]{x} = \left\{\sqrt[3]{x}\right\}^4$$

Similarly, $\sqrt[a]{x^b} = (\sqrt[a]{x})^b$; $\sqrt{x^3} = (\sqrt{x})^3$

In the expression $\sqrt[a]{x^b}$, *if* a *and* b *be both multiplied or divided by the same number, the value of the expression is not altered.* That is,

$$\sqrt[ma]{x^{mb}} = \sqrt[a]{x^b}$$

For it has been shewn that

$$\sqrt[ma]{x^{mb}} = \sqrt[a]{\sqrt[m]{x^{mb}}} \text{ which is } \sqrt[a]{\sqrt[m]{(x^b)^m}} = \sqrt[a]{x^b}$$

$$\text{because} \quad \sqrt[m]{(x^b)^m} = x^b$$

Similarly, $\sqrt[6]{a^4} = \sqrt[3]{a^2} = \sqrt[9]{a^6} = \sqrt[12]{a^8}$

To extract a root of a power, divide the exponent of the power by the index of the root, if that division be possible without fractions. Thus, $\sqrt[4]{x^{12}} = x^{\frac{12}{4}} = x^3$. For $x^{12} = (x^3)^4$; therefore $\sqrt[4]{x^{12}} = x^3$. Similarly, $\sqrt[8]{x^{16}} = x^2$, $\sqrt{x^{20}} = x^{10}$; and so on.

When the exponent of the power is not divisible by the index of the root, as in the case of $\sqrt[3]{x^7}$, we have (at least as yet) no algebraical mode of operation by which to reduce $\sqrt[3]{x^7}$ to any form in which the sign $\sqrt[3]{\ }$ does not appear.* It only remains, therefore, to find what number or fraction x stands for, and then † to calculate $\sqrt[3]{x^7}$ or $x^2\sqrt[3]{x}$ by the rules of arithmetic.

The only question that remains is about a mode of representing $\sqrt[3]{x^7}$; and this question we have anticipated in page 86, where we have found that it would be highly convenient in one respect to let $x^{\frac{1}{2}}$, $x^{\frac{1}{3}}$, &c. stand for $\sqrt{x}$, $\sqrt[3]{x}$, &c. But we stopped our course there, because we had no direct reason to know that all the complicated relations of roots might be obtained from that notation, *without the necessity of applying rules to fractional exponents different from those which are applied to common fractions, or of treating the fractional exponents by rules different from those which apply to whole exponents.* We write in opposite columns instances of the rules which we have proved in the case of whole exponents, and those about which we

* We may proceed as follows. Since $x^7 = x^6.x$, we have $\sqrt[3]{x^7} = \sqrt[3]{x^6 \times x} = \sqrt[3]{x^6} \times \sqrt[3]{x} = x^2\sqrt[3]{x}$; in which, however, the symbol $\sqrt[3]{\ }$ still remains.

† There is a certain distinction to be drawn between the processes of algebra and those of arithmetic. We cannot be properly said to *find* results in algebra, but only to put them into the form in which they can be most easily found by arithmetic. For instance, "What is the sum of a and b?" Of this question $a+b$ is not properly a solution, but a representation, and the proper answer to the question must be deferred till we know what numbers a and b stand for. But in the question, "what is the sum of $8a$ and $5a$?" we can go one step further; for though $8a+5a$ is *an* algebraical representation, it is not the most simple one which the language of the science admits. The latter is $13a$; but we are still without the answer until we know what a stands for. When we come to the step at which we must pass to arithmetic to get any nearer the answer, we shall therefore say we have arrived at the *ultimate* algebraical form. Thus $a+b$ is an ultimate form; $8a+5a$ is not. Again $\sqrt[3]{x^7}$, or at most $x^2\sqrt[3]{x}$, is an ultimate form; $\sqrt[3]{x^{12}}$ is not.

wish to inquire as to fractional exponents; the question being, are the theorems in the second column true, upon the supposition that

$$x^{\frac{m}{n}} \text{ represents } \sqrt[n]{x^m}$$

$$x^6 \times x^4 = x^{6+4} = x^{10}$$

$$\frac{x^6}{x^4} = x^{6-4} = x^2$$

$$(x^6)^2 = x^{6\times 2} = x^{12}$$

$$\sqrt[3]{x^6} = x^{\frac{6}{3}} = x^2$$

$$x^{\frac{2}{3}} \times x^{\frac{1}{2}} = x^{\frac{2}{3}+\frac{1}{2}} = x^{\frac{7}{6}}\,?$$

$$x^{\frac{2}{3}} \div x^{\frac{1}{2}} = x^{\frac{2}{3}-\frac{1}{2}} = x^{\frac{1}{6}}\,?$$

$$(x^{\frac{2}{3}})^{\frac{2}{3}} = x^{\frac{2}{3}\times\frac{2}{3}} = x^{\frac{4}{9}}\,?$$

$$\sqrt[3]{x^{\frac{4}{3}}} = x^{\frac{4}{3}\div 3} = x^{\frac{4}{9}}\,?$$

In the first place we observe that we come by the extension from the last rule in the same way as by others. We have found that when b is divisible by a, $x^{\frac{b}{a}}$ is a correct representation of $\sqrt[a]{x^b}$. But when b is not divisible by a, $x^{\frac{b}{a}}$ has no meaning. We give it a meaning; that is, we say, let it still represent $\sqrt[a]{x^b}$, whatever that may be. We now proceed to investigate the rules which this new symbol requires. The first column is the general case, the second a particular case. The references ( ) are to the pages in which the rules are contained.

What is $x^{\frac{m}{n}} \times x^{\frac{p}{q}}$?

$x^{\frac{m}{n}}$ means $\sqrt[n]{x^m}$

(106) which $= \sqrt[nq]{x^{mq}}$

$x^{\frac{p}{q}}$ means $\sqrt[q]{x^p}$

(106) which $= \sqrt[nq]{x^{np}}$

Therefore $x^{\frac{m}{n}} \times x^{\frac{p}{q}}$

$= \sqrt[nq]{x^{mq}} \times \sqrt[nq]{x^{np}}$

(106) $= \sqrt[nq]{x^{mq} \times x^{np}}$

(84) $= \sqrt[nq]{x^{mq+np}}$

Which is represented by

$x^{\frac{mq+np}{nq}}$

But $\dfrac{mq+np}{nq} = \dfrac{m}{n} + \dfrac{p}{q}$

Therefore $x^{\frac{m}{n}} \times x^{\frac{p}{q}} = x^{\frac{m}{n}+\frac{p}{q}}$

What is $x^{\frac{2}{3}} \times x^{\frac{1}{2}}$?

$x^{\frac{2}{3}}$ means $\sqrt[3]{x^2}$

(106) which $= \sqrt[6]{x^4}$

$x^{\frac{1}{2}}$ means $\sqrt{x}$

(106) which $= \sqrt[6]{x^3}$

Therefore $x^{\frac{2}{3}} \times x^{\frac{1}{2}}$

$= \sqrt[6]{x^4} \times \sqrt[6]{x^3}$

(106) $= \sqrt[6]{x^4 \times x^3}$

(84) $= \sqrt[6]{x^7}$

Which is represented by

$x^{\frac{7}{6}}$

But $\dfrac{7}{6} = \dfrac{2}{3} + \dfrac{1}{2}$

Therefore $x^{\frac{2}{3}} \times x^{\frac{1}{2}} = x^{\frac{2}{3}+\frac{1}{2}}$

The next rule will be more briefly deduced, and without references.

What is $x^{\frac{m}{n}} \div x^{\frac{p}{q}}$?

This is $\sqrt[n]{x^m} \div \sqrt[q]{x^p}$

or $\sqrt[nq]{x^{mq}} \div \sqrt[nq]{x^{np}}$

or $\sqrt[nq]{x^{mq} \div x^{np}}$

or $\sqrt[nq]{x^{mq-np}}$

or $x^{\frac{mq-np}{nq}}$

But $\frac{mq-np}{nq} = \frac{m}{n} - \frac{p}{q}$

Therefore $x^{\frac{m}{n}} \div x^{\frac{p}{q}} = x^{\frac{m}{n} - \frac{p}{q}}$

What is $\left(x^{\frac{m}{n}}\right)^{\frac{p}{q}}$?

This means $\sqrt[q]{\left\{\sqrt[n]{x^m}\right\}^p}$

or (106) $\sqrt[q]{\sqrt[n]{(x^m)^p}}$

or (105) $\sqrt[nq]{x^{mp}}$

or $x^{\frac{mp}{nq}}$

But $\frac{mp}{nq} = \frac{m}{n} \times \frac{p}{q}$

Therefore $\left(x^{\frac{m}{n}}\right)^{\frac{p}{q}} = x^{\frac{m}{n} \times \frac{p}{q}}$

What is $x^{\frac{2}{3}} \div x^{\frac{1}{2}}$?

This is $\sqrt{x^2} \div \sqrt{x}$

or $\sqrt[6]{x^4} \div \sqrt[6]{x^3}$

or $\sqrt[6]{x^4 \div x^3}$

or $\sqrt[6]{x^{4-3}}$

or $x^{\frac{1}{6}}$

But $\frac{1}{6} = \frac{2}{3} - \frac{1}{2}$

Therefore $x^{\frac{2}{3}} \div x^{\frac{1}{2}} = x^{\frac{1}{6}}$

What is $(x^{\frac{4}{5}})^{\frac{2}{3}}$?

This means $\sqrt[3]{\left\{\sqrt[5]{x^4}\right\}^2}$

or (106) $\sqrt[3]{\sqrt[5]{(x^4)^2}}$

or (105) $\sqrt[15]{x^8}$

or $x^{\frac{8}{15}}$

But $\frac{8}{15} = \frac{4}{5} \times \frac{2}{3}$

Therefore $(x^{\frac{4}{5}})^{\frac{2}{3}} = x^{\frac{4}{5} \times \frac{2}{3}}$

The last process contains the answer to both the third and fourth inquiries in page 108.

By looking at page 86, and remembering that the fundamental rules there used have now been proved to be applicable to fractional exponents, the meaning of, and rules relating to, negative fractional exponents may be established. Thus,

$$x^{-\frac{2}{3}} \text{ stands for } \frac{1}{x^{\frac{2}{3}}} \text{ or } \frac{1}{\sqrt[3]{x^2}} \text{ and also for } \sqrt[3]{\left(\frac{1}{x}\right)^2}$$

We shall now proceed to inquire how many *algebraical* roots there may be, and of what kind, in the case of the square, cube, and fourth roots. To go further would be difficult for the beginner at present.

First, as to the square root. It is evident that $+1$ and -1 are both square roots of $+1$, and $+a$ and $-a$ both square roots of $+aa$. For

$$-1 \times -1 = +1 \qquad -a \times -a = +aa$$
$$+1 \times +1 = +1 \qquad +a \times +a = +aa$$

The question now is, can there be more than two square roots to $+1$? Let x be any square root of $+1$, then xx must (by the definition of the term square root) $= 1$, or $xx - 1 = 0$. But $xx - 1 = (x+1)(x-1)$; therefore $(x+1)(x-1) = 0$. Therefore,* either $x+1$ or $x-1$ is $= 0$. To $x+1 = 0$, the only answer is $x = -1$; to $x-1 = 0$, the only answer is $x = +1$; therefore $+1$ and -1 are the only square roots of 1. The same process may be applied to $xx = aa$, or $(x-a)(x+a) = 0$.

Therefore a negative quantity can have no square root which is either a positive or a negative quantity; for either of these, multiplied by itself, is a positive quantity. We will not therefore say that $\sqrt{-1}$ is no "quantity," because we have agreed to give that name to every symbol which results from the rules of calculation. But $\sqrt{-1}$ (whenever it occurs) will be what -1 was in page 47, the evidence of some misconception of a problem, for which the problem must be examined, and altered, extended, or abandoned, as may be found necessary. But if we look at the steps by which we established the meaning of -1, we shall find them to be as follows:†

1. We met with such combinations of symbols as $3-4$, &c. proposing operations which contradicted the meaning which the symbols 3, —, 4, then had.

2. We examined the problems which gave rise to such combinations, and found out how to make the correction without repeating the process; so ascertaining what was to be understood by such expressions as $3-4$, &c. that we could either predict their appearance, or explain them when they appeared.

3. We examined what would arise from applying common rules

* When a product $ab = 0$, either a or $b = 0$, for if both have a value, the product, by common rules, has a value.

† The student may make his understanding what immediately follows a test of his having understood all that precedes.

to $3-4$, &c., and what would be the effect of deferring the correction of the misconception till any later stage of the process.

4. From all that preceded, we extended the meaning of the terms commonly used, and in such a manner, that what were till then simply the results of misconception, became recognised symbols with a definite meaning, and used with demonstrated rules, not differing in practice from those with which they were used in their limited signification.

5. And we found that in all cases in which the result produced was simply arithmetical (that is, consistent and intelligible when the terms had their limited significations), that no error was left in the result by the unintelligible character (with the limited meanings) of the preceding steps; but that the result was the same as it would have been if we had retraced our steps and made each step arithmetical.

Thus we observed that $+$, which before a number, 3, means addition, before -3 is equivalent to a direction to annex -3 to the preceding part of the expression; and that though we could not dispense with the extended meaning of $+$ and $-$ in $+(-6)-(+4)$, yet that $+6+3$ admitted of arithmetical interpretation, even though it were the result of several of the more extended forms of operation.

From this we are at liberty to *conjecture*, that a further extension of the meaning of $+$ and $-$ might, by precisely the same train of operations, give a rational method of using and interpreting $\sqrt{-1}$, $\sqrt{-2}$, &c., which at present are wholly unmeaning and contradictory. We say *conjecture*, because it by no means follows that a method of the success of which we have only one instance, will be universally applicable. Such a method has been given, but it is not our intention to explain it here. We shall simply shew that a field has been left in which the explanation may be looked for.

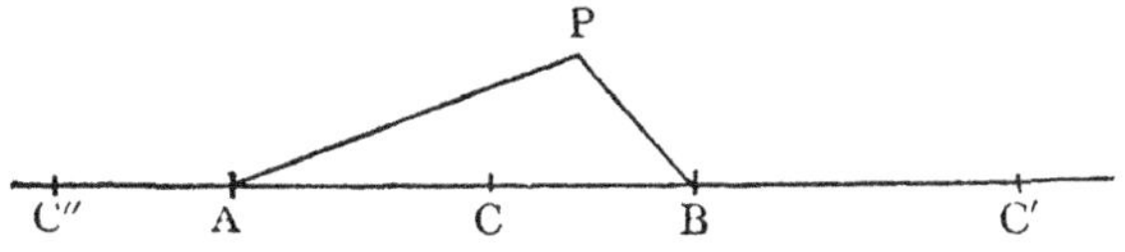

If we suppose a person to set out from A, and stop at B, stopping first at some other point, and always keeping in the line AB or its continuation; and if we suppose distance measured towards the right to be positive, and towards the left negative, we can by our rules ascertain the distance $+AB$ through which he goes altogether from his first position, as follows:

1. If he stop at C $(+AC)+(+CB) = +AB$
2. If he stop at C′ $(+AC')+(-C'B) = +AB$
3. If be stop at C″ $(-AC'')+(+C''B) = +AB$

Now observe that if he should leave the line AB or its continuation, if, for instance, he should go through AP, PB, we have no symbols so connected with AP, &c. that (let ¶ simply denote that some yet uninvented symbol is to be in its place)

$$¶(¶AP)¶(¶PB) = +AB$$

Therefore we find that in the application of algebra we may yet have new symbols to employ, and we also fall upon unexplained symbols such as $\sqrt{-1}$, &c. May not such extensions be made as will make $\sqrt{-1}$, &c, with an extended use of + and —, supply the place of the yet-to-be-invented symbols? This is for the student of this work a point for conjecture only, but it will make the following course advisable.

1. Apply the rules of algebra to such expressions as $\sqrt{-1}$, &c. merely to see what will come of using them, without placing any confidence in the results, or at least more than the experience of a great number shall render unavoidable.

2. Whenever, in the course of a process, it appears that such expressions as $\sqrt{-1}$, &c. have disappeared, examine the result and see whether it is true.

We now pass to the cube root. Let x be any one of the cube roots of 1. Then $x^3=1$ or $x^3-1=0$ $x^3-1=(x-1)(x^2+x+1)=0$ (see page 103, and make $y=1$);

therefore either $x-1=0$ or $x^2+x+1=0$

The first gives $x=1$, and 1 is evidently a cube root of 1. The second cannot be solved till the student has read the next chapter; but its solutions (for it has two) both contain the yet unexplained symbol $\sqrt{-a}$. They are

$$\frac{-1+\sqrt{-3}}{2} \quad \text{and} \quad \frac{-1-\sqrt{-3}}{2}$$

and we shall shew of the first of these (leaving the second to the student), 1. That it does satisfy the equation x^2+x+1, if common rules be considered as applicable to $\sqrt{-3}$; 2. That it is a cube root of 1, on the same supposition.

$$\text{If } x = \frac{-1+\sqrt{-3}}{2} \quad x^2 = \frac{(-1)^2+2(-1)(\sqrt{-3})+(\sqrt{-3})^2}{4}$$

$$= \frac{1-2\sqrt{-3}+(-3)}{4} = \frac{-2-2\sqrt{-3}}{4} = \frac{-1-\sqrt{-3}}{2}$$

Therefore

$$x^2+x+1 = \frac{-1-\sqrt{-3}}{2} + \frac{-1+\sqrt{-3}}{2} + 1 = \frac{-1-1}{2} + 1 = 0$$

$$\text{Again} \qquad x^3 = x^2 \times x = \frac{-1-\sqrt{-3}}{2} \times \frac{-1+\sqrt{-3}}{2}$$

$$= \frac{(-1)^2-(\sqrt{-3})^2}{4} = \frac{1-(-3)}{4} = \frac{4}{4} = 1$$

$$\text{Let} \qquad p = \frac{-1-\sqrt{-3}}{2} \qquad q = \frac{-1+\sqrt{-3}}{2}$$

Then the three cube roots of aaa or a^3 must be a, pa, and qa. For first $a \times a \times a = a^3$.

$$pa \times pa \times pa = p^3a^3 = a^3 \quad \text{because} \quad p^3 = 1$$
$$qa \times qa \times qa = q^3a^3 = a^3 \quad \text{because} \quad q^3 = 1$$

and the same might be deduced from the equation $x^3 - a^3 = 0$, so soon as we know how to solve $x^2 + ax + a^2 = 0$.

Let x be one of the fourth roots of 1. Then we have $x^4 = 1$, or $x^4 - 1 = 0$; that is,

$$(x^2-1)(x^2+1) = 0 \text{ and either } x^2-1 = 0 \text{ or } x^2+1 = 0$$

the solutions of $x^2 - 1 = 0$ are -1 and $+1$, as before; and the solutions of $x^2+1 = 0$ are either $x = +\sqrt{-1}$ or $x = -\sqrt{-1}$. Therefore 1 has four fourth roots, $+1$, -1, $+\sqrt{-1}$, and $-\sqrt{-1}$. This will be found true upon the application of common rules; for instance,

$$(\sqrt{-1})^2 = -1 \quad (\sqrt{-1})^3 = (\sqrt{-1})^2.\sqrt{-1} = -\sqrt{-1}$$
$$(\sqrt{-1})^4 = (\sqrt{-1})^3.\sqrt{-1} = -\sqrt{-1}.\sqrt{-1} =$$
$$-(\sqrt{-1})^2 = -(-1) = 1$$

The only use which we can logically make of such expressions as $\sqrt{-x}$, previous to any further inquiry, is the following: Let a, b, c, and d, be positive or negative quantities. Then

$$a + b\sqrt{-x} = c + d\sqrt{-x}$$

cannot be true unless $a = c$ and $b = d$. For suppose $a = c \pm e$, that is, is different from c, then

$$c \pm e + b\sqrt{-x} = c + d\sqrt{-x} \qquad \sqrt{-x} = \frac{\pm e}{d-b}$$

that is, $\sqrt{-x}$ is a positive or negative quantity, which is absurd. Consequently $(\pm e) \div (d-b)$ cannot be a common algebraical quantity. Now if $e = 0$, and d is not equal to b, we have $\sqrt{-x} = 0 \div (d-b) = 0$, which is not true; if e be finite and $d = b$, we have $\sqrt{-x} = \pm e \div 0$, which does not agree with page 25, since no quantity multiplied by itself is negative because it is great, or nearer to negative the greater the quantity becomes. The only supposition remaining is that $e = 0$ and $d - b = 0$, that is, $a = c$ and $d = b$; and the only form at which we have here arrived for $\sqrt{-x}$ is $\frac{0}{0}$. But this simply indicates that the equation from which it is derived is always true; which is the case (so far as such an equation can yet be said to be true at all), when $a = c$ and $b = d$. Observe that we do not lay much stress on the preceding; it only proves that the equation cannot be true *unless* $a = c$ and $b = d$; but it may be matter of dispute as yet whether the above *is* true if $a = c$ and $b = d$. For we may as yet reasonably refuse our assent even to the equation $x = x$ unless x represent magnitude of some sort; we may say that ideas are contained in the meaning of the sign $=$ which do not apply except to magnitudes. But if any one should say this, we refer him to the extended definition of $=$, page 62, though the beginner must recollect that he never will comprehend the force of that definition as applied to $\sqrt{-x}$, &c. until he has more experience of such symbols.

But there are algebraical quantities analogous to the preceding to which we must now direct attention. They are such as $\sqrt{3}$, $\sqrt{5}$, $\sqrt[3]{2}$, &c. which cannot be exactly found, but only approximately, pages 92, &c. We say that

$$a + b\sqrt{3} = c + d\sqrt{3}$$

cannot be true (if a, b, c, and d be numbers or fractions) unless $a = c$ and $b = d$. The same reasoning applies, word for word, substituting $\sqrt{3}$ for $\sqrt{-x}$, the absurdity deduced being that we cannot have

$$\sqrt{3} = \frac{\pm e}{d-b}$$

where d, b, e, are whole numbers or fractions.

In a similar way it may be proved that if all the letters stand for definite numbers or fractions, and (x and y not being square numbers or fractions)

$$a + b\sqrt{x} = c + d\sqrt{y} \quad \cdots\cdots\cdots \quad (A)$$

Then a must $= c$, and therefore $b\sqrt{x} = d\sqrt{y}$. For if not, let $a = c \pm e$; substitute, and $(-)\ c$

$$\pm e + b\sqrt{x} = d\sqrt{y}$$

Square both sides, which gives

$$(\pm e)^2 + 2(\pm e)b\sqrt{x} + (b\sqrt{x})^2 = (d\sqrt{y})^2$$

or $$e^2 \pm 2eb\sqrt{x} + b^2x = d^2y$$

therefore $$\sqrt{x} = \frac{d^2y - b^2x - e^2}{\pm 2e}$$

that is, $\sqrt{x}$, which cannot be expressed in a definite fraction, is so expressed, which is absurd. The only way of making the equation A possible is, therefore, by making $a = c$ and $b\sqrt{x} = d\sqrt{y}$.

This principle may, in certain cases, be applied to the extraction of the square roots of such quantities as $4 + 2\sqrt{3}$, $21 + 4\sqrt{5}$, &c. Take such a quantity, say $2 + \sqrt{7}$, and square it.

$$(2 + \sqrt{7})^2 = 2^2 + 2 \times 2\sqrt{7} + (\sqrt{7})^2$$
$$= 4 + 4\sqrt{7} + 7 = 11 + 4\sqrt{7}$$

Now, suppose such a quantity as $11 + 4\sqrt{7}$ to be given; how are we to find out, 1. That it has a square root of the same form;* 2. That that square root is $2 + \sqrt{7}$? As follows: if $11 + 4\sqrt{7}$ have a square root of the same form, let it be $x + \sqrt{y}$, so that

$$\sqrt{11 + 4\sqrt{7}} = x + \sqrt{y} \qquad \text{(square both sides)}$$
$$11 + 4\sqrt{7} = x^2 + 2x\sqrt{y} + (\sqrt{y})^2$$
$$= x^2 + y + 2x\sqrt{y}$$

Hence $$x^2 + y = 11 \quad \text{and} \quad 2x\sqrt{y} = 4\sqrt{7}$$

Therefore $$(-) \quad x^2 - 2x\sqrt{y} + y = 11 - 4\sqrt{7}$$

* Observe that there is no essential difference of *form* between $11 + 4\sqrt{7}$ and $x + \sqrt{y}$, for $4\sqrt{7} = \sqrt{(4)^2 \times 7} = \sqrt{112}$; whence $11 + 4\sqrt{7}$ $= 11 + \sqrt{112}$.

but the first side is the square of $x-\sqrt{y}$, or

$$(x-\sqrt{y})^2 = 11-4\sqrt{7} \quad \text{that is, } x-\sqrt{y} = \sqrt{11-4\sqrt{7}}$$

$$\text{But } x+\sqrt{y} = \sqrt{11+4\sqrt{7}}$$

Multiply the last two equations together, which gives

$$(x+\sqrt{y})(x-\sqrt{y}) = \sqrt{11+4\sqrt{7}}\sqrt{11-4\sqrt{7}}$$

$$\text{or } x^2-y=\sqrt{(11+4\sqrt{7})(11-4\sqrt{7})}=\sqrt{121-112}=3$$

$$\text{But } x^2+y= \quad 11$$

$$(+)\quad 2x^2 = 14 \quad x^2 = 7 \quad x = \sqrt{7}$$

$$(-)\quad 2y = 8 \quad y = 4 \quad \sqrt{y} = 2$$

Therefore $\sqrt{11+4\sqrt{7}}$ or $x+\sqrt{y}$ is $\sqrt{7}+2$, as we saw in the method by which $11+4\sqrt{7}$ was obtained.

We shall apply the same process to the formation of $\sqrt{a+b\sqrt{c}}$. Let this be $x+\sqrt{y}$.

Then
$$a+b\sqrt{c} = (x+\sqrt{y})^2$$
$$= x^2+y+2x\sqrt{y}$$

Therefore
$$a = x^2+y \quad \text{and} \quad b\sqrt{c} = 2x\sqrt{y}$$

Therefore
$$a-b\sqrt{c} = x^2+y-2x\sqrt{y} = (x-\sqrt{y})^2$$

or
$$x-\sqrt{y} = \sqrt{a-b\sqrt{c}}$$

But
$$x+\sqrt{y} = \sqrt{a+b\sqrt{c}}$$

$$(\times)\quad x^2-y = \sqrt{(a-b\sqrt{c})(a+b\sqrt{c})} = \sqrt{a^2-b^2c}$$

But
$$x^2+y = \quad a$$

$$(+)\quad 2x^2 = a+\sqrt{a^2-b^2c}$$

$$x = \sqrt{\tfrac{1}{2}a+\tfrac{1}{2}\sqrt{a^2-b^2c}}$$

$$(-)\quad 2y = a-\sqrt{a^2-b^2c}$$

$$\sqrt{y} = \sqrt{\tfrac{1}{2}a-\tfrac{1}{2}\sqrt{a^2-b^2c}}$$

$$x+\sqrt{y} = \sqrt{\tfrac{1}{2}a+\tfrac{1}{2}\sqrt{a^2-b^2c}}+\sqrt{\tfrac{1}{2}a-\tfrac{1}{2}\sqrt{a^2-b^2c}}$$

and this is the square root of $a+b\sqrt{c}$.

Verification. Let $\frac{1}{2}a+\frac{1}{2}\sqrt{a^2-b^2c}$ be called p

Let $\frac{1}{2}a-\frac{1}{2}\sqrt{a^2-b^2c}$ be called q

$$pq = (\tfrac{1}{2}a)^2-(\tfrac{1}{2}\sqrt{a^2-b^2c})^2 = \tfrac{1}{4}a^2-\tfrac{1}{4}(a^2-b^2c)$$

$$= \tfrac{1}{4}a^2-\tfrac{1}{4}a^2+\tfrac{1}{4}b^2c = \tfrac{1}{4}b^2c;\ \text{therefore}\ \sqrt{pq} = \tfrac{1}{2}b\sqrt{c}.$$

According to the preceding theorem,

$$\sqrt{a+b\sqrt{c}} = \sqrt{p}+\sqrt{q} \quad (a+b\sqrt{c}) = (\sqrt{p}+\sqrt{q})^2$$

$$= (\sqrt{p})+2\sqrt{p}\sqrt{q}+(\sqrt{q})^2 = p+2\sqrt{pq}+q$$

But $\quad p+q = \frac{1}{2}a+\frac{1}{2}a = a \quad 2\sqrt{pq} = b\sqrt{c}$

Therefore $p+q+2\sqrt{pq}$ is $a+b\sqrt{c}$; which shews the preceding theorem to be true.

This theorem is of little practical value, but is very good exercise in the use of such expressions as $b\sqrt{c}$, &c. It is a simplification only when a^2-b^2c has a real square root; otherwise, it is the reverse, for a square root of a square root occurs only once in $\sqrt{a+b\sqrt{c}}$, but twice in the value found for it. Thus it simplifies the first of the succeeding expressions, but not the second, though both are equally true.

$$\sqrt{13+2\sqrt{30}} = \sqrt{10}+\sqrt{3}$$

$$\sqrt{13+2\sqrt{31}} = \sqrt{\tfrac{13}{2}+\tfrac{1}{2}\sqrt{45}}+\sqrt{\tfrac{13}{2}-\tfrac{1}{2}\sqrt{45}}$$

Anomaly. Apply the preceding result to a case in which b^2c is greater than a^2, or a^2-b^2c a negative quantity. For example, to $2+\sqrt{8}$ ($a = 2 \quad b = 1 \quad c = 8$),

$$\sqrt{2+\sqrt{8}} = \sqrt{1+\tfrac{1}{2}\sqrt{-4}}+\sqrt{1-\tfrac{1}{2}\sqrt{-4}} \ \ldots. \ (A)$$

Is the square root of $2+\sqrt{8}$, therefore, of the same unexplained character as $\sqrt{-1}$, &c.? Certainly not: for it may be found *quam proximè*, by the rules of arithmetic, lying somewhere between the square roots of $2+\sqrt{4}$ and $2+\sqrt{9}$, or between the square roots of 4 and 5.

Are we then to conclude that the expression (A) is in reality arithmetical? On this we must observe, that a really arithmetical expression may, by rules only, be made to appear impossible. For instance,

$$x + y = (x+c\sqrt{-1})+(y-c\sqrt{-1})$$
$$x^2+y^2 = x^2-(-y^2) = (x)^2-(y\sqrt{-1})^2$$
$$= (x+y\sqrt{-1})\,(x-y\sqrt{-1})$$

To investigate the expression (A) further, extract the square root of each term by the rule.

Let $a = 1 \quad b = \frac{1}{2} \quad c = -4$

$$\sqrt{1+\tfrac{1}{2}\sqrt{-4}} = \sqrt{\tfrac{1}{2}+\tfrac{1}{2}\sqrt{2}}+\sqrt{\tfrac{1}{2}-\tfrac{1}{2}\sqrt{2}}$$
$$\sqrt{1-\tfrac{1}{2}\sqrt{-4}} = \sqrt{\tfrac{1}{2}+\tfrac{1}{2}\sqrt{2}}-\sqrt{\tfrac{1}{2}-\tfrac{1}{2}\sqrt{2}}$$

The second sides of the two preceding equations still contain the square root of a negative quantity; because, since 1 is less than $\sqrt{2}$, $\frac{1}{2}$ is less than $\frac{1}{2}\sqrt{2}$, or $\frac{1}{2}-\frac{1}{2}\sqrt{2}$ is negative. Add the two last:

$$\sqrt{1+\tfrac{1}{2}\sqrt{-4}}+\sqrt{1-\tfrac{1}{2}\sqrt{-4}} = 2\sqrt{\tfrac{1}{2}+\tfrac{1}{2}\sqrt{2}}$$

But this is only the quantity with which we started in another form; for it is

$$\sqrt{4(\tfrac{1}{2}+\tfrac{1}{2}\sqrt{2})} \quad \text{or} \quad \sqrt{2+2\sqrt{2}} \quad \text{or} \quad \sqrt{2+\sqrt{4\times 2}}$$

We have, then, by following an applicable rule, simply committed the inadvertence corresponding to the intentional alteration made in $x+y$ above; and $\sqrt{2+\sqrt{8}}$, or generally $\sqrt{a+b\sqrt{c}}$, if a^2 be less than b^2c, cannot be expressed in a result of the form

$$\sqrt{p+\sqrt{q}}+\sqrt{p-\sqrt{q}}$$

unless q be a negative quantity.

The extraction of $\sqrt[3]{a+b\sqrt{c}}$ is of greater difficulty and less use. We shall, therefore, omit it.

Exercises. Verify the following theorems:

1. The three algebraical cube roots of -1, are -1 and the solutions of the equation $x^2-x+1=0$, which are $\frac{1}{2}+\frac{1}{2}\sqrt{-3}$ and $\frac{1}{2}-\frac{1}{2}\sqrt{-3}$.

2. The four algebraical fourth roots of -1 are

$$\tfrac{1}{2}\sqrt{2}(1+\sqrt{-1}) \qquad \tfrac{1}{2}\sqrt{2}(1-\sqrt{-1})$$
$$\tfrac{1}{2}\sqrt{2}(-1+\sqrt{-1}) \qquad \tfrac{1}{2}\sqrt{2}(-1-\sqrt{-1})$$

3. The eight eighth roots of 1 are

$$\pm 1,\ \pm\sqrt{-1},\ \pm\tfrac{1}{2}\sqrt{2}(1+\sqrt{-1}),\ \pm\tfrac{1}{2}\sqrt{2}(1-\sqrt{-1})$$

Give the reason why this set includes all the values of $\sqrt[4]{-1}$?

Having given a quantity which contains radical* terms of the second degree (that is, square roots) to find a multiplier such that the products shall be free from radicals.

1. A simple radical term, such as $\sqrt{3}$. Here $\sqrt{3}$ is the multiplier, or $a\sqrt{3}$, where a is *rational*;† for $\sqrt{3} \times a\sqrt{3} = 3a$, which is rational.

2. A binomial, one or both terms of which are radical, such as $\sqrt{3}+\sqrt{2}$. Since $(a+b)(a-b) = a^2-b^2$, if a and b are simple radical terms, a^2-b^2 is rational; therefore $a-b$ is the multiplier for $a+b$, and *vice versâ*. For instance, $\sqrt{3}+\sqrt{2}$ multiplied by $\sqrt{3}-\sqrt{2}$ gives 3—2, or 1; $2\sqrt{3}-\frac{1}{2}\sqrt{7}$ multiplied by $2\sqrt{3}+\frac{1}{2}\sqrt{7}$ gives $4 \times 3-\frac{1}{4} \times 7$, or $10\frac{1}{4}$.

3. A trinomial, containing two or three radical terms; such as $\sqrt{3}+\sqrt{5}-\sqrt{7}$ or $\sqrt{a}+\sqrt{b}+\sqrt{c}$. Multiply by $\sqrt{a}+\sqrt{b}-\sqrt{c}$, which gives $(\sqrt{a}+\sqrt{b})^2-(\sqrt{c})^2$ or $a+2\sqrt{ab}+b-c$ or $a+b-c+2\sqrt{ab}$. Multiply now by $a+b-c-2\sqrt{ab}$; which gives $(a+b-c)^2-(2\sqrt{ab})^2$ or $(a+b-c)^2-4ab$. Therefore the multiplier required is the product of $\sqrt{a}+\sqrt{b}-\sqrt{c}$ and $a+b+c-2\sqrt{ab}$.

$$(\sqrt{3}+\sqrt{5}-\sqrt{7})(\sqrt{3}+\sqrt{5}+\sqrt{7}) = 1+2\sqrt{15}$$
$$(1+2\sqrt{15})(1-2\sqrt{15}) = 1-60 = -59$$

It will seldom or never be requisite to consider more than three terms.

The preceding can be applied to find the value of a fraction which has radicals in the denominator. Thus to find $1 \div (\sqrt{3}+1)$ instead of extracting the root of three and forming the denominator, multiply both numerator and denominator by $\sqrt{3}-1$, which gives

$$\frac{1}{\sqrt{3}+1} = \frac{\sqrt{3}-1}{(\sqrt{3}+1)(\sqrt{3}-1)} = \frac{\sqrt{3}-1}{3-1} = \frac{1}{2}(\sqrt{3}-1$$

the second side of the equation is evidently the more easily found :

* *Radix*, Latin for root; *radical* quantities, those which contain roots.

† Rational, a term used in algebra, meaning, free from radicals; for instance, a common number or fraction. Thus 2 is rational, and also $\sqrt{4}$, though in a radical form. But $\sqrt{2}$ is irrational.

$$\frac{\sqrt{6}+\sqrt{7}}{\sqrt{6}-\sqrt{5}}=\frac{(\sqrt{6}+\sqrt{7})(\sqrt{6}+\sqrt{5})}{(\sqrt{6}-\sqrt{5})(\sqrt{6}+\sqrt{5})}=6+\sqrt{42}+\sqrt{30}+\sqrt{35}$$

The only case of higher roots which is worth consideration is where there is a simple radical term, such as $\sqrt[5]{8}$, $a\sqrt[n]{b}$. The multiplier in the first case is $(\sqrt[5]{8})^4$, or $8^{\frac{4}{5}}$ for $8^{\frac{1}{5}}\times 8^{\frac{4}{5}}=8^{\frac{5}{5}}=8$. Thus the following results are obtained :

$$\frac{\sqrt[3]{2}}{\sqrt[3]{3}}=\frac{\sqrt[3]{2}.\sqrt[3]{3}.\sqrt[3]{3}}{\sqrt[3]{3}.\sqrt[3]{3}.\sqrt[3]{3}}=\frac{\sqrt[3]{18}}{3}\,;\quad \frac{\sqrt[5]{3}}{\sqrt[5]{2}}=\frac{\sqrt[5]{48}}{2}$$

$$\frac{\sqrt[n]{a}}{\sqrt[n]{b}} \text{ or } \frac{a^{\frac{1}{n}}}{b^{\frac{1}{n}}}=\frac{a^{\frac{1}{n}}\times b^{\frac{n-1}{n}}}{b^{\frac{1}{n}}\times b^{\frac{n-1}{n}}}=\frac{a^{\frac{1}{n}}\,b^{\frac{n-1}{n}}}{b}$$

$$\frac{1}{q^{\frac{1}{n}}\,r^{\frac{1}{w}}}=\frac{q^{\frac{n-1}{n}}\,r^{\frac{w-1}{w}}}{q\,r}\qquad\qquad \frac{2}{\sqrt[3]{3}\sqrt[4]{4}}=\frac{2\sqrt[3]{9}\sqrt[4]{64}}{12}$$

The following miscellaneous examples on all the matters contained in this chapter should be carefully verified.

$$a\times b^{-1}\times a^{-3}\times b^{\frac{2}{3}}=a^{-2}b^{-\frac{1}{3}}=\frac{1}{a^2b^{\frac{1}{3}}}$$

$$a^7\div a^{-5}=a^{12}=a^7\times a^5=1\div a^{-12}$$

$$a^m\times b^n=b^n\div a^{-m}=a^m\div b^{-n}=1\div a^{-m}b^{-n}$$

$$\sqrt{a\sqrt{a\sqrt{a}}}=a^{\frac{7}{8}}\qquad \sqrt[3]{(a\sqrt{(a\sqrt[3]{a})})}=a^{\frac{5}{9}}$$

$$\sqrt[n]{\left\{\sqrt[m+1]{a^{m-1}b}\right\}^m}=a^{\frac{m^2-m}{mn+n}}\,b^{\frac{m}{mn+n}}$$

$$\left\{\frac{p}{qr^{-1}}\right\}^{\frac{3}{5}}\times\left\{\frac{q}{p^2r}\right\}^{-\frac{1}{3}}=(pq^{-1}r)^{\frac{3}{5}}\times(p^{-2}qr^{-1})^{-\frac{1}{3}}$$

$$=p^{\frac{3}{5}}q^{-\frac{3}{5}}r^{\frac{3}{5}}\times p^{\frac{2}{3}}q^{-\frac{1}{3}}r^{\frac{1}{3}}=p^{\frac{19}{15}}q^{-\frac{14}{15}}r^{\frac{14}{15}}$$

$$=\sqrt[15]{\frac{p^{19}r^{14}}{q^{14}}}=\sqrt[15]{p^{19}}\sqrt[15]{\left(\frac{r}{q}\right)^{14}}=\frac{p^{\frac{19}{15}}q^{\frac{1}{15}}r^{\frac{14}{15}}}{q}$$

$$\sqrt{a^2b^3c}=\sqrt{a^2b^2.bc}=ab\sqrt{bc}=\frac{ab^2c}{\sqrt{bc}}$$

$$\sqrt[3]{a^4b^8c^{13}}=\sqrt[3]{a^3b^6c^{12}.ab^2c}=ab^2c^4\sqrt[3]{ab^2c}$$

$$a^{\frac{4}{3}}b^{\frac{8}{3}}c^{\frac{13}{3}}=a^{1+\frac{1}{3}}b^{2+\frac{2}{3}}c^{4+\frac{1}{3}}=ab^2c^4a^{\frac{1}{3}}b^{\frac{2}{3}}c^{\frac{1}{3}}$$

$\sqrt[\frac{1}{2}]{a} = a^2$ $\sqrt[\frac{1}{3}]{a^2} = a^6$ (never used, but might be, why?)

$$\sqrt{160} = \sqrt{16 \times 10} = 4\sqrt{10}; \quad \sqrt[3]{160} = \sqrt[3]{8 \times 20} = 2\sqrt[3]{20}$$

$$\sqrt{3332} = 14\sqrt{17} \quad \sqrt[3]{3348} = 3\sqrt[3]{124} \quad \sqrt[4]{32} = 2\sqrt[4]{2}$$

$$\sqrt{\frac{3}{7}} = \frac{\sqrt{21}}{7} = \frac{3}{\sqrt{21}} \quad \sqrt{\frac{a}{b}} = \frac{\sqrt{ab}}{b} = \frac{a}{\sqrt{ab}}$$

$$a : \sqrt{ab} :: \sqrt{ab} : b \qquad a^{\frac{1}{3}} : a^{\frac{1}{2}} :: a^{-\frac{1}{3}} : a^{-\frac{1}{6}}$$

$$\frac{1}{\sqrt{\frac{1}{2}} + \sqrt{\frac{1}{3}}} = 6\left(\sqrt{\tfrac{1}{2}} - \sqrt{\tfrac{1}{3}}\right) \qquad \frac{1}{\sqrt{a} - \sqrt{ab}} = \frac{1 + \sqrt{b}}{\sqrt{a}(1-b)}$$

$$\frac{1}{\sqrt{a+1} - 1} = \frac{1}{a}\left(\sqrt{a+1} + 1\right)$$

$$\frac{1}{\sqrt{a+x} + \sqrt{a-x}} = \frac{1}{2x}\left(\sqrt{a+x} - \sqrt{a-x}\right)$$

$$\frac{\sqrt{a+x} + \sqrt{a-x}}{\sqrt{a+x} - \sqrt{a-x}} = \frac{2a + 2\sqrt{a^2 - x^2}}{2x} = \frac{a}{x} + \sqrt{\frac{a^2}{x^2} - 1}$$

$$\sqrt{a^2 - x^2} = a\sqrt{1 - \frac{x^2}{a^2}} = x\sqrt{\frac{a^2}{x^2} - 1} = \sqrt{ax}\sqrt{\frac{a}{x} - \frac{x}{a}}$$

$$\sqrt{a+x} = \sqrt{a}\sqrt{1 + \frac{x}{a}} = \sqrt{x}\sqrt{\frac{a}{x} + 1} = \sqrt{ax}\sqrt{\frac{1}{x} + \frac{1}{a}}$$

$$\tfrac{1}{2}\sqrt{b^2 - 4ac} = \sqrt{\frac{b^2}{4} - ac} = \frac{b}{2}\sqrt{1 - \frac{4ac}{b^2}} = \sqrt{\frac{b^2 - 4ac}{4}}$$

$$\sqrt{2ax - x^2} = x\sqrt{2\frac{a}{x} - 1} = \sqrt{x}\sqrt{2a - x} = a\sqrt{2\frac{x}{a} - \frac{x^2}{a^2}}$$

$$\frac{a + b\sqrt{-1}}{x + y\sqrt{-1}} = \frac{ax + by + (bx - ay)\sqrt{-1}}{x^2 + y^2}$$

$$(1 + \sqrt{-1}) = \sqrt{-1}\,(1 - \sqrt{-1}) \qquad \sqrt{-7} = \sqrt{7}.\sqrt{-1}$$

$$\sqrt{-4} = 2\sqrt{-1} \qquad \sqrt{-\tfrac{1}{2}} = \tfrac{1}{2}\sqrt{2}.\sqrt{-1}$$

$$\sqrt{-\frac{a}{b}} = \frac{\sqrt{ab}}{b}.\sqrt{-1} \quad \sqrt{-4} \times \sqrt{-3} = -\sqrt{12}$$

$$(a - b) = \left(a^{\frac{1}{2}} - b^{\frac{1}{2}}\right)\left(a^{\frac{1}{2}} + b^{\frac{1}{2}}\right) = \left(a^{\frac{1}{3}} - b^{\frac{1}{3}}\right)\left(a^{\frac{2}{3}} + a^{\frac{1}{3}}b^{\frac{1}{3}} + b^{\frac{2}{3}}\right)$$

M

$$= \left(a^{\frac{1}{4}} - b^{\frac{1}{4}}\right)\left(a^{\frac{3}{4}} + a^{\frac{1}{2}}b^{\frac{1}{4}} + a^{\frac{1}{4}}b^{\frac{1}{2}} + b^{\frac{3}{4}}\right)$$

$$a + b = \left(a^{\frac{1}{5}} + b^{\frac{1}{5}}\right)\left(a^{\frac{4}{5}} - a^{\frac{3}{5}}b^{\frac{1}{5}} + a^{\frac{2}{5}}b^{\frac{2}{5}} - a^{\frac{1}{5}}b^{\frac{3}{5}} + b^{\frac{4}{5}}\right)$$

$$\left(a^{\frac{2}{3}} + a^{\frac{1}{3}} + 1\right)^2 = a^{\frac{4}{3}} + 2a + 3a^{\frac{2}{3}} + 2a^{\frac{1}{3}} + 1$$

$$\left(a^{\frac{2}{3}}\right)^{\frac{3}{2}} = a \qquad \left(ab^{\frac{1}{3}}c^{-2}\right)^{-\frac{1}{5}} = a^{-\frac{1}{5}}b^{-\frac{1}{15}}c^{\frac{2}{5}}$$

[N.B. It is usual to call quantities of the forms $a\sqrt{-1}$ or $\sqrt{-a^2}$, $a + \sqrt{-b}$, or $a + \sqrt{b}\sqrt{-1}$, *impossible* quantities. This they are at present, as not having received any interpretation; in the same manner 10—14 was impossible in Chapter I. But, considering that they will in due time (if the student proceed so far) receive their interpretation, though not in this work, we shall call them *purely symbolical.* All the phrases of algebra are symbolical; but all which contain a letter or numeral, which we have yet met with, have an interpretation connected with numbers, making them representatives of magnitude. But $\sqrt{-1}$ has received no such interpretation; it is therefore a pure symbol, as much so as + or —, or more so, inasmuch as yet it indicates neither magnitude nor operation. Hence, in performing operations with pure symbols, we can be guided only by experience, or where that fails, we have new conventions to make. Since it is understood that such symbols will finally be rejected altogether, unless an interpretation present itself which brings them under the dominion of common rules, these rules only should be applied to them. The only case which requires notice is that of forming the symbol which is to represent $\sqrt{-a} \times \sqrt{-b}$. This, by common rules, may be either $\sqrt{-a \times -b}$ or $\sqrt{ab}$, or $\sqrt{a} \times \sqrt{-1}$ multiplied by $\sqrt{b} \times \sqrt{-1}$ or $\sqrt{ab} \times -1$, that is $-\sqrt{ab}$. For reasons hereafter to appear, let the student always take the latter, that is, let

$$\sqrt{-a} \times \sqrt{-b} \text{ be } -\sqrt{ab} \text{ not } +\sqrt{ab}]$$

Having two symbols to indicate the nth root of a, namely, $\sqrt[n]{a}$ and $a^{\frac{1}{n}}$, we shall employ the first in the simple arithmetical sense, and the second to denote any one of the algebraical roots, that is, any one we please, unless some particular root be specified.

Thus $\sqrt{4}$ is 2, without any reference to sign; but $(4)^{\frac{1}{2}}$ may be either $+2$ or -2. Thus $\sqrt[3]{a}$ is the cube root found in arithmetic, while

$$(a)^{\frac{1}{3}} \text{ is either } \sqrt[3]{a}, \quad \frac{-1+\sqrt{-3}}{2}\sqrt[3]{a}, \text{ or } \frac{-1-\sqrt{-3}}{2}\sqrt[3]{a}$$

$$(a)^{\frac{1}{4}} \text{ stands for } \sqrt[4]{a}, \quad -\sqrt[4]{a}, \quad \sqrt{-1}.\sqrt[4]{a}, \text{ and } -\sqrt{-1}\sqrt[4]{a}$$

Similarly $a+b^{\frac{1}{2}}$ has two values, namely, either $a+\sqrt{b}$ or $a-\sqrt{b}$.

Hence $\sqrt{b}$ (when b is positive) always stands for a positive arithmetical quantity. Suppose we wish to represent merely the *numerical* value of b, without reference to its sign; we may abbreviate the following sentence: "the number contained in b, taken positively, whether b be positive or negative," by $\sqrt{b^2}$, which* is the same thing. Thus $b = \pm\sqrt{b^2}$ according as b is positive or negative.

We shall now proceed to the general consideration of expressions of the second degree.

* I have seen $a \pm x$, or what is here signified by $a+(x^2)^{\frac{1}{2}}$, denoted by $a+\sqrt{x^2}$, in which the ambiguity of sign was referred to $\sqrt{}$. But this was only in one place, and though the want of some express stipulation as to the distinction between radical signs and fractional exponents, has led to some variety of usage, I think the conventions in the text will best agree with the majority of writers. At least, though I do not pretend to have made any research expressly on this point, $\pm\sqrt{a}$ is very familiar to my eye, and $\pm a^{\frac{1}{2}}$ not at all so.

CHAPTER V.

GENERAL THEORY OF EXPRESSIONS OF THE FIRST AND SECOND DEGREES; INCLUDING THE NUMERICAL SOLUTION OF EQUATIONS OF THE SECOND DEGREE.

THE view taken in the First Chapter of equations of the first degree simply amounted to their *numerical solution;* that is, having given two expressions not higher than the first degree with respect to x, required that value of x which will make the two expressions equal. We there saw that all equations of the first degree could be reduced to others of the form $ax = b$; thus, in page 3, we reduced

$$\frac{x}{2} + \frac{x}{3} = 1 - \frac{x}{4} \quad \text{to} \quad 13x = 12$$

It is most convenient to bring all the terms of an equation on one side; thus $ax - b = 0$ is in a more convenient form for investigation than $ax = b$. In this manner the whole theory of equations is considered as involving two fundamental inquiries. The first is;—*Having given an algebraical expression which contains* x, *required that value, or those values of* x, *which make the expression vanish, that is, become equal to* 0.

We here (in compliance with common custom) use the term *root*, in a manner different from its use in the last chapter. Every value of x which makes an expression containing x equal to 0, or, as the phrase is, makes it *vanish*, is called* a *root* of that expression. Thus,

* This might be considered an extension of the former meaning, as follows. Let *root* have the meaning above assigned, then the square root of 3 (of last chapter) is the root (as here defined) of x^2-3, the cube root of 3 is the root of x^3-3; and so on. But we do not make this as an extension, because there is no corresponding extension for the correlative term *power*. And we must confess, that we use this new meaning of the term *root* with some repugnance, in spite of its shortness and convenience. Would not the *nullifier* do as well? If we were inventing algebraical terms anew, we should certainly say that 7 is the *nullifier* of $2x-14$, and that the latter is *nullified* when $x = 7$. But it is not advisable to introduce words or sybmols which the student will not afterwards see in the best writers on the applications of mathematics.

in page 2, we used language which we may now modify as follows: "The root of $2x-1-5x+19$ is 6;" "the roots of $16x-x^2-48$ are 4 and 12;" "the roots of $x^3-6x^2+11x-6$ are 1, 2, and 3."

The second fundamental inquiry is as follows:—*Having given an algebraical expression which contains* x, *what values of* x *make that expression positive, what values make it negative, what values make it purely symbolical* (See page 122)? We shall proceed to answer these questions for expressions of the first degree. We first make the following remarks.

1. As we wish the student to keep in mind that we consider various values of x, and the consequences deduced from them as to the sign of the expression, we shall (whenever we may think it necessary) use some modification of this latter to denote the root. Thus, instead of saying that $2x-14$ vanishes when $x=7$, or that $x=7$ is the root, we shall call the root $x_{,}$; the second root, if there be one, $x_{,,}$, and so on.

2. We shall always suppose that the co-efficients, a, b, &c. in such expressions as $ax+b$, ax^2+bx+c, &c. are either positive or negative algebraical quantities; unless the contrary be specially mentioned. Thus, we shall treat of cases where a is $1, 2, -1, -\frac{1}{2}\sqrt{3}$, &c. but never, without special mention, of the case where a is $\sqrt{-1}$, or $1+\sqrt{-3}$, &c. *But we make no such limitation with respect to* x.

3. We shall suppose the student to be familiar with the use of the transformations in page 73; for instance, that he can make $ax+b$ coincide with $2x-3$ by writing the latter as $2x+(-3)$, and supposing $a=2, b=-3$.

The general form of the expression of the first degree containing x is $ax+b$. For instance

$$\frac{x-5}{2}-\frac{x-2}{3}+x \text{ is } \frac{x}{2}-\frac{5}{2}-\frac{x}{3}+\frac{2}{3}+x$$

which is $\left(\frac{1}{2}-\frac{1}{3}+1\right)x-\left(\frac{5}{2}-\frac{2}{3}\right)$

which coincides with $ax+b$, by supposing

$$a=\frac{1}{2}-\frac{1}{3}+1 \quad b=-\left(\frac{5}{2}-\frac{2}{3}\right)$$

The root of $ax+b$ is readily found; for let

$$ax+b=0, \text{ then } x=-\frac{b}{a} \text{ call this } x_{,}.$$

When $ax+b$ vanishes, it may be written $ax_{,}+b$, and $x_{,}$ denotes $-\frac{b}{a}$. From $x_{,}=-\frac{b}{a}$ we get $ax_{,}=-b$ or $b=-ax_{,}$. Write this value instead of b in $ax+b$, which then becomes $ax-ax_{,}$ or $a(x-x_{,})$. Consequently we have this THEOREM. *If $x_{,}$ be the root of $ax+b$, then*

$$ax+b=a(x-x_{,}) \text{ for all values of } x.$$

The last is not an *equation of condition* (See introduction) but an *identical* equation. It implies that the two sides are absolutely the same, but in different forms: indeed it may be thus immediately traced from $ax+b$ without any alteration except of form,

$$ax+b=a\left(x+\frac{b}{a}\right)=a\left\{x-\left(-\frac{b}{a}\right)\right\}=a(x-x_{,})$$

because it has been laid down that $x_{,}$ stands for $-\frac{b}{a}$.

The preceding will seem an unnecessary repetition, but, on coming to expressions of the second degree, the reason of it will be seen.

EXERCISE. Point out, by inspection, the roots of the following

expressions	$3x+\frac{1}{2}$,	$-4x-3$,	$\frac{1}{2}x-\frac{2}{3}$
roots	$-\frac{\frac{1}{2}}{3}$	$-\frac{-3}{-4}$	$-\frac{-\frac{2}{3}}{\frac{1}{2}}$
reduced roots	$-\frac{1}{6}$	$-\frac{3}{4}$	$+\frac{4}{3}$
altered expressions	$3\left\{x-\left(-\frac{1}{6}\right)\right\}$,	$-4\left\{x-\left(-\frac{3}{4}\right)\right\}$,	$\frac{1}{2}\left\{x-\frac{4}{3}\right\}$

THEOREM. *The expression* a x + b *is of the same sign as* a, *when* x *is greater than the root, and of a different sign from* a *when* x *is less than the root.*

For, $ax+b$ is $a(x-x_{,})$; if x be greater than $x_{,}$, $x-x_{,}$ is positive, (page 63), and $a\times$ *positive quantity*, retains the sign of a (page 64); but if x be less than $x_{,}$, $x-x_{,}$ is negative, and $a\times$ *negative quantity* changes the sign of a.

EXAMPLES. $3x+\frac{1}{2}$ is positive for every value of x greater than $-\frac{1}{6}$; for instance for $-\frac{1}{8}$: this we shall try. If $x=-\frac{1}{8}$

$$3x+\frac{1}{2}=3\times-\frac{1}{8}+\frac{1}{2}=+\frac{1}{2}-\frac{3}{8}=+\frac{1}{8}.$$

The same is negative for every value of x less than $-\frac{1}{6}$; for example, for $x=-\frac{1}{5}$. For then

$$3x+\frac{1}{2}=3\times-\frac{1}{5}+\frac{1}{2}=\frac{1}{2}-\frac{3}{5}=-\frac{1}{10}.$$

Similarly $-4x-3$ is negative (that is, the same as -4) for every value of x greater than $-\frac{3}{4}$, and positive for every value of x less than $-\frac{3}{4}$: but $\frac{1}{2}x-\frac{2}{3}$ is positive for every value of x greater than $\frac{4}{3}$; and negative for every value of x less than $\frac{4}{3}$.

This is sufficient to render our notions of expressions of the first degree consistent with what we are now going to lay down concerning those of the second.

Lemma 1. If $x+y=p+q$, and $xy=pq$, then x is $=$ one of the two, p or q, and y is $=$ the other.

Square the first equation, and from the result subtract the second multiplied by 4, as follows,

$$\begin{array}{rcl} x^2+2xy+y^2 & = & p^2+2pq+q^2 \\ 4xy & = & 4pq \\ \hline (-) \quad x^2-2xy+y^2 & = & p^2-2pq+q^2 \end{array}$$

The square root of the first side is either of the two, $x-y$ or $y-x$; that of the second $p-q$ or $q-p$. Extract the square root, which gives therefore one of the four following equations:

$$x-y=p-q \cdots\cdot (1) \qquad y-x=p-q \cdots\cdot (2)$$
$$x-y=q-p \cdots\cdot (3) \qquad y-x=q-p \cdots\cdot (4)$$

But $x+y=p+q$; which last combined with the four preceding, separately, gives as follows:

with (1) or (4) $x=p$ $y=q$ } which was to
with (2) or (3) $x=q$ $y=p$ } be shewn.

(This lemma will certainly seem most superfluous to the student: but he must recollect that though, "if $x=p$ or q, and $y=q$ or p, it is most evident that $x+y=p+q$, and $xy=pq$," yet that the converse, namely, that "if $x+y=p+q$ and $xy=pq$, then x *cannot be any thing but* p or q, and y *cannot be any thing but* q or p," is not equally evident. Thus if

$$x^2-2ax = b^2-2ab$$

it is most evident that $x = b$ satisfies this equation, but by no means evident that nothing but $x = b$ satisfies it. In fact $x = 2a-b$ will be also found to satisfy it.)

Lemma 2. The product of two expressions of the first degree [say $ax+b$ and $a'x+b'$] *cannot be* always *equal to the product of any other two expressions of the first degree* [say $cx+e$ and $c'x+e'$] *unless the two latter be made from the two former by multiplying and dividing by a quantity independent of* x [that is, unless $ax+b = m(cx+e)$ and $a'x+b' = \frac{1}{m}(c'x+e')$ where m is independent of x]. For

$$(ax+b)(a'x+b') = aa'x^2+(ab'+a'b)x+bb' \cdots\cdots (A)$$
$$(cx+e)(c'x+e') = cc'x^2+(ce'+c'e)x+ee' \cdots\cdots (B)$$

If possible, let these two developements* be *always* equal, whatever value is given to x. That is, let

$$px^2+qx+r = p'x^2+q'x+r'$$

where p stands for aa', p' for cc', &c. (This is merely for abbreviation.) Now, these two cannot be always equal unless they are absolutely identical, that is, unless $p = p'$ $q = q'$ and $r = r'$. This we prove as follows:—If the two sides of the preceding equation be always equal, they are equal when $x = 1$, and also when $x = 2$, and also when $x = 3$. Let t_1 t_2 t_3† be the values of the first side of the equation when x is successively made 1, 2, and 3. Then the other sides will have the same values, according to our supposition; that is,

when $x = 1$ $\quad p+q+r = t_1 \quad p'+q'+r' = t_1$
when $x = 2$ $\quad 4p+2q+r = t_2 \quad 4p'+2q'+r' = t_2$
when $x = 3$ $\quad 9p+3q+r = t_3 \quad 9p'+3q'+r' = t_3$

Now, apply the first set of equations to find p, q, and r, supposing

* Developement, any expression formed by giving another expression a more expanded form.

† These are distinct quantities; for the like abbreviation see page 103.

t_1 t_2 and t_3 known, as is done in page 80. Then apply the second set to find p', q', and r'. It is clear that, from the perfect likeness of the equations, p, q, and r, are found by exactly the same operations on the same quantities which give p', q', and r'. Consequently, the results will be the same, or we shall have $p = p'$ $q = q'$ $r = r'$. As an exercise, we give the three results of the first set, which are

$$p = \frac{t_3 - 2t_2 + t_1}{2} \quad q = \frac{8t_2 - 3t_3 - 5t_1}{2} \quad r = t_3 - 3t_2 + 3t_1$$

[The following proof is more simple; but it involves the supposition of a letter being made equal to 0, which we do not wish to use till after the discussion in a succeeding chapter.

If $\quad px^2 + qx + r = p'x^2 + q'x + r'$ *always*,

it is true, among other cases, when $x = 0$; but it then is reduced to

$$0 + 0 + r = 0 + 0 + r' \quad \text{or} \quad r = r'$$

Therefore $\quad px^2 + qx + r = p'x^2 + q'x + r \quad$ always

$(-)r \quad px^2 + qx \quad = p'x^2 + q'x \quad \cdots\cdots$

$(\div)x \quad px + q \quad = p'x + q' \quad \cdots\cdots$

This must also be true when $x = 0$, or

$$0 + q = 0 + q' \quad \text{that is} \quad q = q'$$

Therefore $\quad px + q = p'x + q \quad (-)q \quad px = p'x$ or $p = p'$]

It has been proved, then, that the expressions (A) and (B), in page 128, cannot be always equal, whatever x may be, unless

$$aa' = cc', \quad ab' + a'b = ce' + c'e, \quad \text{and} \quad bb' = ee'$$

Divide the second and third by the first, which gives

$$\frac{ab'}{aa'} + \frac{a'b}{aa'} = \frac{ce'}{cc'} + \frac{c'e}{cc'}, \qquad \frac{bb'}{aa'} = \frac{ee'}{cc'},$$

or $$\frac{b'}{a'} + \frac{b}{a} = \frac{e'}{c'} + \frac{e}{c}, \qquad \frac{b'}{a'} \times \frac{b}{a} = \frac{e'}{c'} \times \frac{e}{c}$$

whence, by Lemma 1, either $\frac{b'}{a'} = \frac{e'}{c'}$, $\frac{b}{a} = \frac{e}{c}$, or $\frac{b'}{a'} = \frac{e}{c}$, and $\frac{b}{a} = \frac{e'}{c'}$.

Suppose the first.

But $\quad ax + b = a\left(x + \frac{b}{a}\right) = a\left(x + \frac{e}{c}\right)$

$$= a \cdot \frac{cx + e}{c} = \frac{a}{c}(cx + e)$$

Similarly $a'x + b' = \frac{a'}{c'}(c'x + e')$

But since $\frac{aa'}{cc'} = 1$ or $\frac{a}{c} \times \frac{a'}{c'} = 1$, if $\frac{a}{c} = m$, $\frac{a'}{c'} = \frac{1}{m}$

Therefore $$ax + b = m(cx + e)$$

$$a'x + b' = \frac{1}{m}(c'x + e')$$

If the second assumption be taken, let the student shew that a similar result will be obtained.

To proceed with the numerical solution of equations of the second degree, we shall take the most simple algebraical form, $ax^2 + bx + c$, to which any other expression of the second degree may be reduced (page 125). Thus $-\frac{1}{2}x^2 + 2x - 3$ agrees with it, if $a = -\frac{1}{2}$ $b = 2$ $c = -3$.

Definition. An expression is said to be a perfect square *with respect to x*, when its square root can be extracted in a form which does not shew x under the sign $\sqrt{}$. Thus, as may be found by trial,

$$\sqrt{ax^2 + 2a^2x + a^3} = \sqrt{a}(x + a)$$
$$\sqrt{a^2x + 2ax^2 + x^3} = \sqrt{x}(a + x)$$

Hence $ax^2 + 2a^2x + a^3$ is a perfect square with respect to x, but not with respect to a; while $a^2x + 2ax^2 + x^3$ is a perfect square with respect to a, but not with respect to x. Observe, that if $px^2 + qx + r$ be a perfect square with respect to x, it remains so after multiplication by any quantity which does not contain x, for if

$$px^2 + qx + r = (gx + h)^2$$

then $$mpx^2 + mqx + mr = \left\{\sqrt{m}(gx + h)\right\}^2$$

Lemma. The condition which implies that $px^2 + qx + r$ is a perfect square with respect to x, is $q^2 = 4pr$.

Let us suppose that $gx + h$ is the square root of $px^2 + qx + r$ (no other form can be, as may be proved by trial). Then

$$(gx + h)^2 = px^2 + qx + r$$

or $$g^2x^2 + 2ghx + h^2 = px^2 + qx + r \qquad \text{(always)}$$

Therefore (page 128)

$$g^2 = p, \qquad 2gh = q, \qquad h^2 = r$$

Here, then, are *three* equations to determine *two* (yet) undetermined quantities g and h. If the product of the first and third be multiplied by 4, and if the second be squared, we have

$$4g^2h^2 = 4pr \quad \text{and} \quad (2gh)^2 \quad \text{or} \quad 4g^2h^2 = q^2$$

Therefore $q^2 = 4pr$, which condition must be satisfied, if the three equations between g and h are to be true.

The preceding equations give $g = \sqrt{p}$, $h = \sqrt{q}$, or $\sqrt{p}x + \sqrt{q}$ is (if $q^2 = 4pr$) the square root of $px^2 + qx + r$.

Some ambiguity may here arise from $\sqrt{p}$ and $\sqrt{q}$, as (from page 110) $g^2 = p$ gives g either $+\sqrt{p}$ or $-\sqrt{p}$, and similarly h is either $+\sqrt{q}$ or $-\sqrt{q}$.

But here observe, that the equation $q^2 = 4pr$ was obtained (partly) by transforming $2gh = q$ into $4g^2h^2 = q^2$. But the latter might also have been obtained in the same way from $2gh = -q$, in which $-q$ is written in place of q; consequently, the same equation implies that both the following are perfect squares, $px^2 + qx + r$ and $px^2 - qx + r$. And if we take the two values of g and h, and combine them in every possible way in the expression $gx + h$, we shall have the four following expressions :—

$$\sqrt{p}x + \sqrt{r} \qquad\qquad -\sqrt{p}x + \sqrt{r}$$
$$\sqrt{p}x - \sqrt{r} \qquad\qquad -\sqrt{p}x - \sqrt{r}$$

each of which is either a square root of $px^2 + qx + r$ or of $px^2 - qx + r$. But, returning to the untransformed equation $2gh = q$, which belongs to the former expression only, as $2gh = -q$ does to the latter, (both being represented in $4g^2h^2 = q^2$) we see that all the four values of $gx + h$ cannot be square roots of $px^2 + qx + r$, but only those in which g and h have such signs, that the product gh may have the same sign as q. For instance, if q be positive, g and h are either both positive or both negative; because gh must in that case be positive : if q be negative, either g is positive and h negative, or g negative and h positive.

Thus, if q be positive, and $q^2 = 4pr$, the square roots of $px^2 + qx + r$ are $\sqrt{p}x + \sqrt{r}$, and $-\sqrt{p}x - \sqrt{r}$; if q be negative the roots of $px^2 - qx + r$ are $\sqrt{p}x - \sqrt{r}$ and $-\sqrt{p}x + \sqrt{r}$. These agree with page 110, where it appears that the two square roots of a quantity differ only in sign; for

$$-\sqrt{p}\,x-\sqrt{r}=-(\sqrt{p}\,x+\sqrt{r})$$
$$-\sqrt{p}\,x+\sqrt{r}=-(\sqrt{p}\,x-\sqrt{r})$$

EXAMPLES. $3x^2+2x+1$ is not a square, because $(2)^2$, or 4, is not equal to $4\,(3\times 1)$, or 12; but $2x^2-12x+18$ is a square, because $(-12)^2$, or 144, is $=4\times(2\times 18)$, or 144. Its square root is either $\sqrt{2}x-\sqrt{18}$ (q is negative) or $-\sqrt{2}x+\sqrt{18}$.

The principal use of the preceding theorem is to *complete a square*, as it is called; that is, to supply either of the terms px^2, qx, or r, by means of the other two. For instance, to make $2x^2+3x$ a complete square. Here $p=2$, $q=3$, r is not given. But that the above may be a square we must have $q^2=4pr$, that is $9=8r$, or $r=\frac{9}{8}$, and we find that $2x^2+3x+\frac{9}{8}$ is a complete square, its roots being

either $$\sqrt{2}x+\frac{3}{\sqrt{8}} \quad\text{or}\quad -\sqrt{2}x-\frac{3}{\sqrt{8}}.$$

Generally, if $q^2=4pr$, $r=\frac{q^2}{4p}$, so that px^2+qx is made a complete square by the addition of

$$\frac{(\text{co-efficient of } x)^2}{4\ (\text{co-efficient of } x^2)}$$

Thus, $ax^2+bx+\frac{b^2}{4a}$ is a perfect square, and so is $4a^2x^2+4abx+b^2$, the roots of the latter being $\pm(2ax+b)$.

We now proceed to distinguish the peculiarities of different forms of the expression

$$ax^2+bx+c$$

If $b^2=4ac$, we have seen that the expression is a complete square. We shall then look separately at the cases in which b^2 is greater than $4ac$, and in which b^2 is less than $4ac$.

1. Let b^2 be greater than $4ac$, or let*

$$b^2=4ac+e^2 \quad\text{or}\quad 4ac=b^2-e^2$$

* Why $4ac+e^2$ rather than $4ac+e$? Because we wish to signify that $4ac$ is really increased. In $4ac+e$ we do not know whether there is increase or decrease, till we know whether e is positive or negative (page 63). But e^2 is positive, whether e be positive or negative (purely symbolical quantities being out of the question). Hence the form of a square is a convenient method by which the student may bear in mind that a quantity is positive.

Now $ax^2+bx+c=\frac{4a^2x^2+4abx+4ac}{4a}=\frac{4a^2x^2+4abx+b^2-e^2}{4a}$

$$=\frac{(2ax+b)^2-e^2}{4a}=\frac{(2ax+b+e)(2ax+b-e)}{4a}$$

or, we have this theorem:

If $\quad e^2=b^2-4ac \qquad$ or $\quad e=\sqrt{b^2-4ac}$

$$ax^2+bx+c=\frac{1}{4a}(2ax+b+e)(2ax+b-e)$$

the two being identically equal.

2. Let $b^2=4ac$, then ax^2+bx+c is a perfect square, and so is $4a^2x^2+4abx+4ac$, which is $4a^2x^2+4abx+b^2$; and

$$ax^2+bx+c=\frac{4a^2x^2+4abx+b^2}{4a}=\frac{(2ax+b)^2}{4a}$$

3. Let b^2 be less than $4ac$, that is, let

$$b^2=4ac-e^2 \qquad \text{or} \quad 4ac=b^2+e^2$$

$$ax^2+bx+c=\frac{4a^2x^2+4abx+4ac}{4a}=\frac{4a^2x^2+4abx+b^2+e^2}{4a}$$

$$=\frac{(2ax+b)^2+e^2}{4a}$$

Previously to proceeding further, we shall apply the preceding expressions to particular cases.

1. Let the expression be $3x^2-7x+4$. Here $a=3$, $b=-7$, $c=4$. And $b^2=49$, $4ac=48$, whence b^2 is greater than $4ac$, and $b^2-4ac=1$. This is e^2; therefore $e=+1$, or -1. Let $e=+1$

$$3x^2-7x+4=\frac{(6x-7+1)(6x-7-1)}{4\times 3}=\frac{(6x-6)(6x-8)}{12}$$

$$=\frac{6(x-1)\times 2(3x-4)}{12}=(x-1)(3x-4)$$

as may easily be verified by multiplication. Let the student shew that the supposition of $e=-1$ gives the same result.

$$\begin{array}{r} 3x-4 \\ x-1 \\ \hline 3x^2-4x \qquad \\ -3x+4 \\ \hline 3x^2-7x+4 \end{array}$$

Now, we ask, what are the roots of this expression, or the values of x which make it vanish. A product becomes $=0$ if either of its factors becomes $=0$; that is, let $x-1=0$, or, let $3x-4=0$, and in the first case

$$3x^2-7x+4 = 0\times(3-4) = 0:$$

in the second, $3x^2-7x+4 = \left(\frac{4}{3}-1\right)\times 0 = 0$

But if $x-1=0$, $x=1$, and if $3x-4=0$, $x=\frac{4}{3}$, therefore 1 and $\frac{4}{3}$ are the values of x which make $3x^2-7x+4$ vanish, or its *roots* (page 124).

We now inquire what values of x will make

$$3x^2-7x+4 \quad \text{or its equal} \quad (x-1)(3x-4)$$

positive or negative. It appears that $x-1$ is positive when x is greater than 1, and $3x-4$ when x is greater than $\frac{4}{3}$; while the first is negative when x is less than 1, and the second when x is less than $\frac{4}{3}$, (page 126).

Value of x.	Sign of $x-1$.	Sign of $3x-4$.	Sign of the product $(x-1)(3x-4)$
Less than 1	$-$	$-$	$+$
Greater than 1, less than $\frac{4}{3}$	$+$	$-$	$-$
Greater than $\frac{4}{3}$	$+$	$+$	$+$

It appears, then, that the preceding expression is always positive, except when x lies between the roots 1 and $\frac{4}{3}$. In this manner we have determined the following points with regard to $3x^2-7x+4$: it is $+$ when x is greater than $\frac{4}{3}$; 0 when $x=\frac{4}{3}$; $-$ when x is less than $\frac{4}{3}$, and greater than 1; 0 when x is 1; $+$ when x is less than 1. Follow a similar process with the following expressions.

$$2x^2+3x+1 = (x+1)(2x+1)$$
$$3x^2+4x-7 = (x-1)(3x+7)$$
$$-2x^2+6x-4 = (2-x)(2x-2)$$

Hitherto we have chosen expressions containing no irrational results: let us now try $3x^2+5x-1$. Here we have $a=3$, $b=5$, $c=-1$; b^2 or 25 is greater (page 62) than $4ac$ or -12, and

$b^2 - 4ac = 37 = e^2$; therefore $e = \pm \sqrt{37}$. Let e be $+\sqrt{37}$, then, page 133,

$$3x^2+5x-1 = \frac{(2\times 3x+5+\sqrt{37})(2\times 3x+5-\sqrt{37})}{4\times 3}$$

$$= \frac{1}{12}(6x+5+\sqrt{37})(6x+5-\sqrt{37})$$

Hence the roots of the expression, which call $x_{,}$ and $x_{,,}$ are

$$x_{,} = -\frac{\sqrt{37}+5}{6} \qquad x_{,,} = \frac{\sqrt{37}-5}{6}$$

$$= -1\cdot8471271 \qquad = \cdot1804604 \text{ very nearly.}$$

It will be found, as before, that the preceding expression is never negative, except when x lies between $x_{,}$ and $x_{,,}$.

2. Let us take $3x^2 - 6x + 3$. Here b^2 or $(-6)^2$ is equal to $4ac$, or $4\times 3\times 3$. Hence the expression is a perfect square, and we have (page 131)

$$3x^2-6x+3 = (\sqrt{3}x-\sqrt{3})^2 = 3(x-1)^2$$

This expression vanishes only when $\sqrt{3}x - \sqrt{3}$ vanishes, or when $x = 1$. But, because there are two equal factors, each of which is $\sqrt{3}x - \sqrt{3}$, and to preserve analogy with the preceding case, it is *said* to have *two* roots, which are *equal*. Thus this expression has two roots, each $= 1$.

This expression is never negative, for $(x-1)^2$ is positive in all cases. We can only make it negative by giving a purely symbolical value to x: for example, $1+\sqrt{-1}$. Then $(x-1)^2$ (by rules only, see page 122) will be -1.

3. In no case hitherto taken has b^2 been less than $4ac$. Now try $2x^2 - x + 4$. Here $a=2$, $b=-1$, $c=4$; and b^2 is 1, $4ac$ is 32, greater than b^2. Here, as in page 133, let $4ac - b^2 = e^2$, which is therefore 31.

Hence, page 133, we have

$$2x^2-x+4 = \frac{(2\times 2x-1)^2+31}{4\times 2} = \frac{(4x-1)^2+31}{8}$$

This expression has no positive or negative root, for $(4x-1)^2$ being always positive, so long as x is positive or negative, must increase 31, and, therefore, $(4x-1)^2+31$ can never $=0$, but is always positive. We see, then, that $2x^2-x+4$ is always positive, for every positive or negative value of x. The least value of the expression under this limitation is $\frac{31}{8}$, for the least value of $(4x-1)^2$ is found by making $4x-1=0$, or $x=\frac{1}{4}$. Consequently, the above expression has the following property: its least value is $\frac{31}{8}$, found by making $x=\frac{1}{4}$; for every other value of x it is greater.

The following cases may also be tried by the student.

$$x^2+x+1 = \frac{1}{4}\left\{(2x+1)^2+3\right\}$$

$$x^2-x+1 = \frac{1}{4}\left\{(2x-1)^2+3\right\}$$

$$-2x^2+2x-5 = -\frac{1}{8}\left\{(4x-2)^2+36\right\}$$

We can give the preceding expressions purely symbolical roots; for instance, to make $2x^2-x+4=0$, let us make

$$(4x-1)^2+31=0 \qquad (4x-1)^2=-31$$

$$4x-1=+\sqrt{-31} \quad \text{or} \quad 4x-1=-\sqrt{-31}$$

Call the roots derived from these $x_{,}$ and $x_{,,}$

$$x_{,}=\frac{1+\sqrt{-31}}{4} \qquad x_{,,}=\frac{1-\sqrt{-31}}{4}$$

which will be found to be roots, by rules only, as in page 122.

We shall now take the more general cases.

1. $ax^2+bx+c=0$, where $b^2-4ac=e^2$ (page 132)

and $ax^2+bx+c=\frac{1}{4a}(2ax+b+e)(2ax+b-e)$

The expression ax^2+bx+c contains eight different forms, as follows, which we shall distinguish by the eight letters A, B, C, D, A′, B′, C′, D′.

	Sign of a.	Sign of b.	Sign of c.
(A) $2x^2+5x+1$	+	+	+
(A′) $-2x^2-5x-1$	−	−	−
(B) $2x^2-5x+1$	+	−	+
(B′) $-2x^2+5x-1$	−	+	−
(C) $2x^2+5x-1$	+	+	−
(C′) $-2x^2-5x+1$	−	−	+
(D) $2x^2-5x-1$	+	−	−
(D′) $-2x^2+5x+1$	−	+	+

We shall first consider that which is common to all the forms; and then the peculiarities of each.

The roots of $ax^2 + bx + c$ are found by solving the following equations (call $x_{,}$ and $x_{,,}$ their roots)

$$2ax + b - e = 0 \qquad 2ax + b + e = 0$$

$$x_{,} = \frac{-b+e}{2a} \qquad x_{,,} = \frac{-b-e}{2a}$$

But $e = \sqrt{b^2 - 4ac}$, therefore

$$x_{,} = \frac{-b + \sqrt{b^2 - 4ac}}{2a} \qquad x_{,,} = \frac{-b - \sqrt{b^2 - 4ac}}{2a}$$

And, page 126, the expression $2ax + b - e$ is the same as $2a(x - x_{,})$ and $2ax + b - e$ as $2a(x - x_{,,})$. This may also be shewn again, thus,

$$2ax + b - e = 2a\left(x + \frac{b-e}{2a}\right) = 2a\left(x - \frac{-b+e}{2a}\right) = 2a\left(x - x_{,}\right)$$

$$2ax + b + e = 2a\left(x + \frac{b+e}{2a}\right) = 2a\left(x - \frac{-b-e}{2a}\right) = 2a\left(x - x_{,,}\right)$$

$$\therefore ax^2 + bx + c = \frac{4a^2 \cdot (x - x_{,})(x - x_{,,})}{4a} = a(x - x_{,})(x - x_{,,})$$

Hence, when the two roots of an expression of the second degree are known, and the coefficient of its first term, the expression itself is known. For instance, what is the expression whose roots are 2 and $-\frac{1}{2}$, and the coefficient of whose first term is 4? By the preceding formula, this expression must be

$$4(x-2)\left(x - (-\tfrac{1}{2})\right) \text{ or } 4(x-2)(x + \tfrac{1}{2}) \text{ or } 4x^2 - 6x - 4$$

If we develope the preceding expression, we find

$$a(x - x_{,})(x - x_{,,}) = ax^2 - a(x_{,} + x_{,,})x + ax_{,}x_{,,}$$

which is identical with $ax^2 + bx + c$; therefore, page 128, we have

$$b = -a(x_{,} + x_{,,}) \text{ or } x_{,} + x_{,,} = -\frac{b}{a}$$

$$c = ax_{,}x_{,,} \qquad \text{or } x_{,}x_{,,} = \frac{c}{a}$$

$$\text{Sum of the roots} = -\frac{\text{Coefficient of } x}{\text{Coefficient of } x^2}$$

$$\text{Product of the roots} = \frac{\text{Term independent of } x}{\text{Coefficient of } x^2}$$

When $a = 1$, or the expression is $x^2 + bx + c$, we have

Sum of the roots $= -b$ Product of the roots $= c$

Verify this theorem on all the preceding examples.

We shall denote the preceding forms by placing the signs of the terms in brackets: thus A will be denoted by $(+ + +)$, A′ by $(- - -)$, &c. This first case, in which $b^2 - 4ac$ is positive, includes all possible varieties of

$$(+ + -) \quad (- - +) \quad (+ - -) \quad (- + +)$$

for in all these, a and c have different signs; ac is negative, and, therefore, $b^2 - 4ac$ is positive and greater than b^2. Here then, $b^2 - 4ac$ is positive without reference to the numerical value of a, b, and c. This same case may or may not include the following,

$$(+ + +) \quad (- - -) \quad (+ - +) \quad (- + -)$$

in all of which ac is positive and therefore the sign of $b^2 - 4ac$ depends upon the simple arithmetical magnitudes of b^2 and ac.

We shall now examine the cases which have roots; and remark that either of the expressions in any one pair may be reduced to the other, by simple change of sign. Thus $-x^2 - x + 1 = -(x^2 + x - 1)$ or $(- - +)$ becomes $(+ + -)$ by entire change of signs only. And, since, when $A = 0$, then $-A = 0$, the expressions in the first of the following columns have roots similar to the corresponding expressions in the second, in every circumstance which depends only upon the signs of the terms.

$(+ + +)$	$(- - -)$	$(+ + -)$	$(- - +)$
$(+ - +)$	$(- + -)$	$(+ - -)$	$(- + +)$

I. *Expressions* $(+ + +)$ $(- - -)$; roots not necessarily existing; and $b^2 - 4ac$ less than b^2. The roots of this expression (when it has them) are both negative. For, since $b^2 - 4ac$ is less than b^2, $\sqrt{b^2 - 4ac}$ is less than $\sqrt{b^2}$, or, page 123, the numerical value of b. Therefore $-b + \sqrt{b^2 - 4ac}$ and $-b - \sqrt{b^2 - 4ac}$ have* the same sign as $-b$, or a contrary sign to b. But the roots are

$$\frac{-b + \sqrt{b^2 - 4ac}}{2a} \qquad \frac{-b - \sqrt{b^2 - 4ac}}{2a}$$

* For instance, both values of -3 ± 2 must be negative; -3 ± 6 has one positive and one negative value.

both of which, as to sign, are therefore negative, because b and a have the same signs in both, and the numerator is of a contrary sign to, and the denominator of the same sign as, a and b.

II. *Expressions* $(+-+)$ $(-+-)$; roots not necessarily existing; and b^2-4ac less than b^2. By a process exactly similar to the preceding, remembering that a and b have now different signs, we prove that both the roots, when they exist, are *positive*.

III. *Expressions* $(++-)$ $(--+)$; roots always existing; b^2-4ac greater than b^2. These two expressions have one positive and one negative root, *the negative root being* numerically *the greater*. For in this case,* since $\sqrt{b^2-4ac}$ is numerically greater than b, $-b+\sqrt{b^2-4ac}$ and $-b-\sqrt{b^2-4ac}$ have different signs; namely, the first is $+$, and the second $-$. Therefore,

$$\frac{-b+\sqrt{b^2-4ac}\text{ (which is +)}}{2a} \text{ agrees in sign with } a \text{ and } b.$$

$$\frac{-b-\sqrt{b^2-4ac}\text{ (which is }-)}{2a} \text{ differs in sign from } a \text{ and } b.$$

If a and b be positive, the second (which is then $-$) is numerically the greater (by the preceding note): if a and b be negative, the first (which is then $-$) is numerically the greater. Therefore in both cases the negative root is numerically the greater.

IV. *Expressions* $(+--)$ $(-++)$; roots always existing; b^2-4ac greater than b^2. Here, by reasoning precisely similar, it may be proved that there must be one positive and one negative root; but that *the positive root is* numerically *the greater*. Observe that a and b have here different signs.

In all these cases we have also the following theorem. *The expression* ax^2+bx+c, when it has different roots, *never differs in sign from* a, *except when the value of* x *lies between that of the roots*. (Read page 134 over again, with attention.) For we have

$$ax^2+bx+c \text{ always } = a(x-x_{,})(x-x_{,,})$$

One of the two roots $x_{,}$ and $x_{,,}$ must be the greater; let it be $x_{,}$. Then, if x be greater than $x_{,}$, it is greater than $x_{,,}$; and $x-x_{,}$ and

* Remember that in $p+q$ it is the sign of that which is *numerically* the greater, which determines the sign of the expression; and that in $p\pm q$ that one, either $p+q$ or $p-q$, is numerically the greater, in which both terms $+p$ and $\pm q$ have the same sign.

$x-x_{,,}$ are both positive. Therefore $a(x-x_{,})(x-x_{,,})$ has the same sign as a. Let x be less than $x_{,}$, but greater than $x_{,,}$ (that is, let x lie between the two roots), then $x-x_{,}$ is negative, $x-x_{,,}$ is positive; and $a(x-x_{,})(x-x_{,,})$ differs in sign from a. Let x be less than $x_{,,}$, then it is less than $x_{,}$; and $x-x_{,}$ and $x-x_{,,}$ are both negative; therefore $a(x-x_{,})(x-x_{,,})$ has the same sign as a. A recapitulation of these three cases gives the theorem in question.

2. $ax^2+bx+c=0$ where $b^2=4ac$ or $b^2-4ac=0$

This case requires that a and c should have the same sign, because $4ac$ must be positive.

Here $$ax^2+bx+c=\frac{(2ax+b)^2}{4a}$$

The two equal roots are derived from

$$2ax+b=0 \quad \text{or} \quad x_{,}=x_{,,}=-\frac{b}{2a}$$

which are positive when b and a differ in sign, that is, in $(+-+)$ and $(-+-)$; and negative when b and a agree in sign, that is in $(+++)$ and $(---)$. The other cases are entirely excluded, since a and c must have the same sign.

The expression ax^2+bx+c being always a square (a positive quantity) divided by $4a$, always has the same sign as a; observe that x cannot now lie *between the roots*.

3. $ax^2+bx+c=0$ $\qquad 4ac-b^2=e^2$ (page 133)

Here a and c must have the same sign, because $4ac$ is positive, being b^2+e^2, the sum of two positive quantities.

(Page 133) $$ax^2+bx+c=\frac{1}{4a}\left\{(2ax+b)^2+e^2\right\}$$

and being a positive quantity divided by $4a$, always has the same sign as a.

The purely symbolical roots (see page 136) are derived from the equation.

$$(2ax+b)^2+e^2=0 \quad \text{or} \quad (2ax+b)^2=e^2\times -1$$

or* $$2ax+b=\pm e\sqrt{-1}=\pm\sqrt{4ac-b^2}\sqrt{-1}$$

* Observe that $p^2=q^2$ or $\pm p=\pm q$, gives only two distinct forms; for $+p=+q$ and $-p=-q$ are the same, as also are $+p=-q$ and $-p=+q$.

$$x_{,} = \frac{-b+\sqrt{4ac-b^2}\sqrt{-1}}{2a} \qquad x_{,,} = \frac{-b-\sqrt{4ac-b^2}\sqrt{-1}}{2a}$$

These roots, using rules only, will be found to satisfy the equation, and also the equations

$$x_{,}+x_{,,} = -\frac{b}{a} \qquad x_{,}x_{,,} = \frac{c}{a}$$

but we cannot at present make an extension of the theorem in page 139, because we can attach no notion of *greater* or *less* to $x_{,}$ and $x_{,,}$.

The numerical solution of equations of the second degree is usually performed by a *process* instead of a *formula*, each case by itself, as follows:

Let $$2x^2-7x+3 = 0$$

(−)3 $$2x^2-7x = -3$$

(÷)2 $$x^2-\frac{7}{2}x = -\frac{3}{2}$$

Complete* the square, $x^2-\frac{7}{2}x+\left(\frac{7}{4}\right)^2 = \left(\frac{7}{4}\right)^2-\frac{3}{2} = \frac{25}{16}$

Extract the root, $x-\frac{7}{4} = \pm\frac{5}{4}$

$$x = \frac{7}{4}+\frac{5}{4} \text{ or } 3; \qquad \text{or } x = \frac{7}{4}-\frac{5}{4} \text{ or } \frac{1}{2}$$

But we should, by all means, desire the student to commit the following theorem to memory:

If $$ax^2+bx+c = 0$$

$$x = \text{either } \frac{-b+\sqrt{b^2-4ac}}{2a} \text{ or } \frac{-b-\sqrt{b^2-4ac}}{2a}$$

EXAMPLES. 1. What are the solutions of

$$px^2+q^2x = qx^2-p^2x+p^3$$

or $$(p-q)x^2+(p^2+q^2)x-p^3 = 0$$

Here $a = p-q \qquad b = p^2+q^2 \qquad c = -p^3$

The roots are the two values of the expression

$$\frac{-(p^2+q^2)\pm\sqrt{(p^2+q^2)^2-4(p-q)(-p^3)}}{2(p-q)}$$

* See page 132, where it appears that $x^2+bx+\frac{b^2}{4}$ is a perfect square, namely, that of $x+\frac{b}{2}$.

$$(p^2+q^2)^2 = p^4+2p^2q^2+q^4$$
$$-4(p-q)(-p^3) = 4p^4-4p^3q$$

Therefore the roots are contained in

$$\frac{-(p^2+q^2)\pm\sqrt{5p^4-4p^3q+2p^2q^2+q^4}}{2(p-q)}$$

2. Let $\quad ax^2-abx = b^2x-b^3$

$$ax^2-(ab+b^2)x+b^3=0$$

The roots are contained in

$$\frac{ab+b^2\pm\sqrt{(ab+b^2)^2-4ab^3}}{2a}$$

$$(ab+b^2)^2-4ab^3 = a^2b^2+2ab^3+b^4-4ab^3$$
$$= a^2b^2-2ab^3+b^4 = (ab-b^2)^2$$

Therefore the roots are contained in

$$\frac{ab+b^2\pm(ab-b^2)}{2a}$$

But $\quad \dfrac{ab+b^2+ab-b^2}{2a} = \dfrac{2ab}{2a} = b$ one root,

$\dfrac{ab+b^2-ab+b^2}{2a} = \dfrac{2b^2}{2a} = \dfrac{b^2}{a}$ the other root.

Verification, $\quad b+\dfrac{b^2}{a} = \dfrac{ab+b^2}{a} = -\dfrac{-(ab+b^2)}{a}$

$b\times\dfrac{b^2}{a} = \dfrac{b^3}{a}$ } (page 137.)

The student should now proceed as follows:

1. To form examples of numerical equations;—choose two roots and a coefficient for the first term, and construct the expression which should have those roots, as in page 137; then find the roots of the resulting expression by the preceding formula, which should be, of course, the roots first chosen. Afterwards take any expressions at hazard; find their roots, and verify them by actual substitution.

2. To construct literal expressions which shall afford solutions of more interest than those taken at hazard, choose any expression which is identically $=0$, in which one letter has no higher power than the second; such as

$$ab^2-abc+abc-ab^2=0$$

write x instead of b in such places as will create an expression, of which it could not be known at first sight that it is made to vanish by $x = b$. For instance, suppose

$$ax^2 - abc + acx - abx = 0$$

one of the roots of this should be b. Find the roots by the formula.

Or take the following method: Choose any two simple expressions, one of which only has a denominator; such as $\frac{m}{n}$ and p. Then the roots of

$$nx^2 - (m+np)x + mp = 0$$

should be $\frac{m}{n}$ and p. For instance, take b and $\frac{1-ab}{a}$. Then

$$m = 1-ab \quad n = a \quad p = b$$

$$m+np = 1-ab+ab = 1; \quad mp = b-ab^2$$

Therefore the roots of

$$ax^2 - x + b - ab^2 = 0$$

should be $\quad x_{,} = b$ and $x_{,,} = \frac{1-ab}{a}$

Anomaly. In the expression $ax^2+bx+c = 0$ let $a = 0$. It then becomes $bx+c = 0$, giving $x = -\frac{c}{b}$. But if we examine the roots of $ax^2+bx+c = 0$ upon the supposition that $a = 0$, we find

$\frac{-b+\sqrt{b^2-4ac}}{2a}$ assumes the form $\frac{0}{0}$ (page 25.)

$\frac{-b-\sqrt{b^2-4ac}}{2a}$ assumes the form $\frac{-2b}{0}$ (page 21.)

Are we then to say, in conformity to the pages cited, that one root is *infinite*, and the other what we please? Apparently not, in the present case; we must therefore examine it further. Instead of supposing $a = 0$, let us (page 21), suppose it as small as may hereafter be necessary. The Lemma which we here lay down will be useful in every part of algebra.

Lemma. The expression $\sqrt{b^2+v}$, may, by supposing v sufficiently small, be made to differ from b by as small a quantity as we please: and, moreover, the same expression may, under the same circumstances, be made to differ from $b+\frac{v}{2b}$, not only by as small a quantity as we please, but by as small a fraction of v as we please.

[In explanation, $\sqrt{1+v}$, however small v may be taken, exceeds 1 by something near the half of v; but $\sqrt{1+v}$ may be made to differ from $1+\frac{1}{2}v$ by less than the ten-millionth part of v, if necessary.]

The first part of this lemma is evident enough: the second we prove as follows:

$$\left(b+\frac{v}{2b}+\sqrt{b^2+v}\right)\left(b+\frac{v}{2b}-\sqrt{b^2+v}\right)$$

$$=\left(b+\frac{v}{2b}\right)^2-\left(b^2+v\right)$$

$$=b^2+2b\times\frac{v}{2b}+\frac{v^2}{4b^2}-b^2-v$$

$$=b^2+v+\frac{v^2}{4b^2}-b^2-v=\frac{v^2}{4b^2}$$

Therefore $$b+\frac{v}{2b}-\sqrt{b^2+v}=\frac{\frac{v^2}{4b^2}}{b+\frac{v}{2b}+\sqrt{b^2+v}}$$

But the last-mentioned fraction has a denominator, which, when v is diminished, approaches continually to $b+\sqrt{b^2}$ or $2b$. Let it be called $2b+w$, where, by making v as small as may be necessary, we can make w as small as we please. Then will

$$b+\frac{v}{2b}-\sqrt{b^2+v}=\frac{v^2}{4b^2(2b+w)}=\frac{v}{4b^2(2b+w)}\times v$$

that is, $b+\frac{v}{2b}$ differs from $\sqrt{b^2+v}$ by a certain fraction of v, namely, $\frac{v}{4b^2(2b+w)}$ of v. But since v can be made as small as we please, and thence w (see what comes before), that is, since $4b^2(2b+w)$ can be brought as near as we please to $4b^2\times 2b$ or $8b^3$, the fraction of v, by which $b+\frac{1}{2v}$ differs from $\sqrt{b^2+v}$, may be thus represented:

$$\frac{v\text{ (a quantity as small as we please)}}{8b^3\text{ (a given quantity)}+\text{ a quantity as small as we please}}$$

and may therefore be made as small as we please.

Precisely the same sort of demonstration may be given of the following; namely, that $b-\frac{v}{2b}$ may be made to differ from $\sqrt{b^2-v}$ by as small a fraction of v as we please.

We shall now proceed to apply this theorem to the consideration of the roots

$$\frac{-b+\sqrt{b^2-4ac}}{2a} \quad \text{and} \quad \frac{-b-\sqrt{b^2-4ac}}{2a}$$

on the supposition that a may be as small as we please. Hence, c being a given quantity, $4ac$ may be as small as we please; and if $v = 4ac$ (by the last lemma)

$\sqrt{b^2-4ac}$ may be made to differ from $b-\frac{4ac}{2b}$ by as small a fraction of $4ac$ as we please.

Let, therefore, $\sqrt{b^2-4ac} = b-\frac{4ac}{2b}-p\times 4ac$, in which we may make p (with a) as small as we please.

Then the roots are

$$\frac{-b+b-\frac{4ac}{2b}-4pac}{2a} \quad \text{and} \quad \frac{-b-b+\frac{4ac}{2b}+4pac}{2a}$$

or $$-\frac{c}{b}-2pc \quad \text{and} \quad \frac{-2b+\frac{2ac}{b}+4pac}{2a}$$

Now, diminish a more and more, in which case p is diminished in the same way. The first root continually approximates to $-\frac{c}{b}$, and the second to the form $-\frac{2b}{0}$. But the first is the root derived from the earlier view of the equation $ax^2+bx+c=0$ in the case where $a=0$, namely, $bx+c=0$, which gives $x=-\frac{c}{b}$. The second is yet unexplained.

Problem in Illustration. a, b, c, and e, are four numbers, the last three of which are increased by a certain number, and the first by m times that number. The results are then found to be proportionals. What is the number?

Let x stand for the number. Then $mx+a$, $x+b$, $x+c$ and $x+e$ are proportionals.

or $$\frac{mx+a}{x+b} = \frac{x+c}{x+e} \quad \text{or} \quad (mx+a)(x+e) = (x+b)(x+c)$$

Perform these multiplications, and reduce the result to an equation of the form $P=0$, which gives

$$(m-1)x^2+(me+a-b-c)x+ae-bc = 0$$

The values of x are contained in

$$\frac{-(me+a-b-c)\pm\sqrt{(me+a-b-c)^2-4(m-1)(ae-bc)}}{2(m-1)}$$

and therefore, generally speaking, there are two solutions of the problem. But if $m=1$, that is, if x must be so chosen that $x+a$, $x+b$, $x+c$, and $x+d$ are proportionals, the case we wish to consider arises: for $m-1=0$; the equation is reduced to

$$(e+a-b-c)x+ae-bc=0$$

which gives only one root; and one of the roots just given takes the form

$$\frac{-2(e+a-b-c)}{0}$$

The interpretation of this form in page 25 was, that any very great number would nearly satisfy the conditions of the problem, a still greater number still more nearly, and so on. Now, the question becomes, will $x+a$, $x+b$, $x+c$, $x+e$, approach more and more nearly to proportionals as x is increased; that is, will

$$\frac{x+a}{x+b}=\frac{x+c}{x+e} \text{ approach to truth in that case?}$$

Divide both numerator and denominator of both fractions by x; this gives $\dfrac{1+\frac{a}{x}}{1+\frac{b}{x}}=\dfrac{1+\frac{c}{x}}{1+\frac{e}{x}}$ which may be made as near the truth as we please, by taking x sufficiently great; for, by so doing, $\frac{a}{x}$, $\frac{b}{x}$, $\frac{c}{x}$, and $\frac{d}{x}$, may be made as small as we please, and the preceding equatio brought as near to $\frac{1}{1}=\frac{1}{1}$ as we please.

Hence it appears, that when a problem which, generally speaking, has two solutions, has a particular case in which there is only one, we may say that there is another solution corresponding to an *infinite* value of the unknown quantity, in the sense explained in page 25.

But, though we see a confirmation of the interpretation put upon $\frac{-2b}{0}$ in page 25, we also see that $\frac{0}{0}$, which is the form of the other root, does not admit the interpretation of page 25, namely, that any value of x will satisfy the equation; but it indicates that the rational root is $-\frac{c}{b}$. We shall return to this point in the next chapter.

The case of $a=0$ presents new circumstances: let us now suppose only $c=0$. We have then $ax^2+bx=0$, or $x(ax+b)=0$; which is satisfied either by $x=0$ or by $ax+b=0$. That is, the roots are 0 and $-\frac{b}{a}$. This would also appear from the general expressions for the roots.

Similarly, if $b=0$, we have $ax^2+c=0$.

$$x^2=-\frac{c}{a} \qquad x=+\sqrt{-\frac{c}{a}} \text{ or } -\sqrt{-\frac{c}{a}}$$

which pair consists of a positive and negative quantity when c and a have different signs, and is purely symbolical when c and a have similar signs. This also would follow directly from the general expressions.

We choose one from among many instances of the use to which the preceding theory may be put. Suppose we know the sum of two quantities (s), and their product (p). Required expressions involving nothing but this sum and product, which shall give the sum of the squares, or cubes, or fourth powers, &c. of the two quantities.

By page 138, these two quantities are the roots of the expression

$$x^2-sx+p=0 \quad (\times)x^n \quad x^{n+2}-sx^{n+1}+px^n=0$$

Represent the roots by $x_{,}$ and $x_{,,}$; we have then

$$x_{,}^{n+2}-sx_{,}^{n+1}+px_{,}^{n}=0$$
$$x_{,,}^{n+2}-sx_{,,}^{n+1}+px_{,,}^{n}=0$$

$$(+) \qquad x_{,}^{n+2}+x_{,,}^{n+2}-s\left(x_{,}^{n+1}+x_{,,}^{n+1}\right)+p\left(x_{,}^{n}+x_{,,}^{n}\right)=0$$

Let the sum of the nth powers of $x_{,}$ and $x_{,,}$ be called A_n; the preceding then becomes

$$A_{n+2}-sA_{n+1}+pA_n=0$$

or $$A_{n+2}=sA_{n+1}-pA_n$$

Now $$A_0=x_{,}^{0}+x_{,,}^{0}=1+1=2 \text{ (page 85)}$$
$$A_1=x_{,}+x_{,,}=s$$

Therefore, by the preceding equation,

$$A_2=sA_1-pA_0= \quad s^2-2p$$
$$A_3=sA_2-pA_1=s(s^2-2p)-ps=s^3-3ps$$
$$A_4=sA_3-pA_2=s(s^3-3ps)-p(s^2-2p)=s^4-4ps^2+2p^2$$

and so on.

There are many equations which may be solved by the assistance of the preceding theory: the reason being, that though they are not strictly of the second degree with respect to the unknown quantity, they are so with respect to some expression containing it. For instance, we wish to find the values of x which satisfy,

$$x^2-3x+1 = 2-\sqrt{x^2-3x+1}$$

In order to clear this equation of the radical sign, we should proceed as follows.

$$\sqrt{x^2-3x+1} = 1+3x-x^2; \text{ square both sides,}$$
$$x^2-3x+1 = 1+6x+7x^2-6x^3+x^4$$
or
$$x^4-6x^3+6x^2+9x = 0$$

an equation of the fourth degree, for which no method of solution has preceded. But, on looking at the original equation, we see immediately that it is of the form $v^2=2-v$; for if $\sqrt{x^2-3x+1}$ be v, then x^2-3x+1 is v^2. Let $v=\sqrt{x^2-3x+1}$, then

$$v^2+v-2=0 \qquad v=1 \text{ or } -2$$

First, let $v=1$

$$\sqrt{x^2-3x+1}=1 \text{ or } x^2-3x+1=1 \therefore x \text{ is } 0 \text{ or } 3$$

Next, let $v=-2$

$$\sqrt{x^2-3x+1}=-2 \quad x^2-3x+1=4 \quad x=\frac{3\pm\sqrt{21}}{2}$$

Therefore the preceding equation is satisfied by the following values of x;

$$0 \qquad 3 \qquad \frac{3+\sqrt{21}}{2} \qquad \frac{3-\sqrt{21}}{2}$$

Again, suppose $2x^6-3=x^3$. Here x^6 is $(x^3)^2$; let $v=x^3$, and the equation becomes $2v^2-3=v$, the roots of which are -1 and $\frac{3}{2}$. Hence, $x^3=-1$ or $x^3=\frac{3}{2}$; that is, any values, real or purely symbolical, of $\sqrt[3]{-1}$ and $\sqrt[3]{\frac{3}{2}}$ are roots of $2x^6-3=x^3$.

We shall close this chapter with some instances of the process of clearing an equation of the radical sign. Let it be the following

$$\sqrt{x}+\sqrt{x+1}+\sqrt{x+2}=2$$
$$\therefore \quad \sqrt{x+\sqrt{x+1}}=2-\sqrt{x+2}$$

Square both sides

$$x+2\sqrt{x}.\sqrt{x+1}+x+1 = 4-4\sqrt{x+2}+x+2$$

or
$$2\sqrt{x(x+1)}+4\sqrt{x+2} = 5-x$$

Square both sides again,

$$4x(x+1)+16\sqrt{x(x+1)}\sqrt{x+2}+16(x+2) = 25-10x+x^2$$

or
$$16\sqrt{x(x+1)(x+2)} = -(7+30x+3x^2)$$

Square both sides again,

$$256x(x+1)(x+2) = (7+30x+3x^2)^2$$

which contains no radical sign, and may be developed.

The following are more simple instances, which we leave to the student.

The equation $\sqrt{x+5}+\sqrt{x-3} = 4$

gives $x-4 = 0$

The equation $\sqrt{x+a}+\sqrt{x+b} = c$

gives $4c^2x+4ab-(c^2-a-b)^2 = 0$

But we must observe that

$$(x+a)^{\frac{1}{2}}+(x+b)^{\frac{1}{2}} = c$$ (see page 123).

gives the same result as the last, and admits of the four following forms:

$\sqrt{x+a}+\sqrt{x+b} = c$	$-\sqrt{x+a}+\sqrt{x+b} = c$
$\sqrt{x+a}-\sqrt{x+b} = c$	$-\sqrt{x+a}-\sqrt{x+b} =$

The value of x above obtained, namely

$$\frac{(c^2-a-b)^2-4ab}{4c^2}$$

will only satisfy one of these. Consequently, when we obtain *one* of the preceding equations, we cannot be sure but that the problem has been misunderstood and requires an extension of form which will give *another* of the preceding.

We give the following as exercises:

1. Shew that $a+\frac{1}{a}$ cannot be *numerically* less than 2. Prove this by shewing that the roots of $a+\frac{1}{a}=2-p$ are purely symbolical

when p lies between 4 and 0. It may also be shewn from $(a-1)^2$ being always positive.

2. Shew that a^2+b^2 must be greater than $2ab$.

3. Prove that if $a+\frac{1}{a}=s$,

$$a^2+\frac{1}{a^2}=s^2-2 \quad a^3+\frac{1}{a^3}=s^3-3s \quad a^4+\frac{1}{a^4}=s^4-4s^2+2$$

4. If $x_{,}$ and $x_{,,}$ be the roots of the expression ax^2+bx+c, then will

$$x_{,}-x_{,,}=\pm\frac{1}{a}\sqrt{b^2-4ac} \quad \frac{x_{,}}{x_{,,}}=\frac{b^2-2ac}{2ac}\pm\frac{b}{2ac}\sqrt{b^2-4ac}$$

$$\frac{x_{,}}{x_{,,}}+\frac{x_{,,}}{x_{,}}=\frac{b^2-2ac}{ac} \qquad \frac{1}{x_{,,}}+\frac{1}{x_{,}}=-\frac{b}{c}$$

5. In the expression ax^2+bx+c, supposing it previously known that one root exceeds the other by m, find the roots without the assistance of the formula. Do the same on the supposition that one root is n times the other.

CHAPTER VI.

ON LIMITS AND VARIABLE QUANTITIES.

We have already had occasion to observe the effects of particular suppositions, which make what in other cases are intelligible quantities, assume the forms $\frac{c}{0}, \frac{0}{0}, a^0$, &c. To these we shall now add the form 0 itself, as requiring investigation on account of the circumstances under which we may have to use it; for since we have found it convenient to reduce every equation to the form $P = 0$, we might, without proper caution, be led to such inferences as the following: If $ab = 0$ and $ac = 0$, then $ab = ac$ or $b = c$. [We* shall give a striking instance of this, as follows: If $x - 2 = 0$, it follows that $x^2 - 4 = 0$, and that $x^2 - 2x = 0$. Are we then to equate $x^2 - 4$ and $x^2 - 2x$, and proceed in any manner, previously explained, with the results? If we do so, since $x^2 - 4 = (x-2)(x+2)$ and $x^2 - 2x = x(x-2)$, we have

$$(x-2)(x+2) = (x-2)x \qquad (\div)\,(x-2) \qquad x+2 = x$$

$$\text{But} \qquad x-2 = 0 \qquad \therefore\ x = 2 \qquad \text{or} \qquad 4 = 2$$

an absurd result, which indicates some absurdity in the process. The suspicious step is the division of both sides of an equation by $x - 2$, which is 0. If we go through the preceding process without the concealment of 0 which takes place by making the supposition $x - 2 = 0$ and then using $x - 2$ instead of 0, we shall find a most evident fallacy, amounting to the following:

$$a \times 0 = 0 \quad b \times 0 = 0 \quad \therefore\ a \times 0 = b \times 0 \quad (\div)\,0 \ a = b$$

that is, we have used 0 as a quantity, have asserted $0 = 0$, and have divided by 0. Returning now to the principle in page 21, we shall suppose $x - 2$ instead of being $= 0$, to be *very small*, and shall put in

* All that comes between the brackets [ ] is vaguely stated; the object being nothing more than to shew the student how liable he is to error in using such terms as *nothing*, *small*, *great*, *nearly equal*, &c.

opposite columns the two analogous processes, each containing its own error.

Let $x-2=0$	Let $x-2$ be as small as we please
$\therefore\ x^2-4=0$	$\therefore\ x^2-4$ may be as small as we please
and $x^2-2x=0$	and x^2-2x may be as small as we please
$\therefore\ x^2-2x=x^2-4$	$\therefore\ x^2-2x$ and x^2-4 may be as nearly equal as we please
or $x(x-2)=(x-2)(x+2)$	and the same of $x(x-2)$ and $(x-2)(x+2)$
$\div(x-2)\quad x=x+2$	$(\div)(x-2)$, then x and $x+2$ may be as nearly equal as we please.
But $x-2=0\quad x=2$	But x may be as near to 2 as we please; therefore 2 and 4 may be as nearly equal as we please.
$\therefore\quad 2=4$	

In the second column there is an error, whichever of the senses in page 24 is put upon the word *equal*. If we call quantities nearly equal whose difference is small, then we do not know that because x^2-2x and x^2-4 are nearly equal, they will still be so after division by $x-2$. For if $x-2$ were, for instance, $\frac{1}{1000}$, then division by $x-2$ is multiplication by 1000, or the difference between the quotients x and $x+2$ is 1000 times the difference of x^2-4x and x^2-4. And the less $x-2$ is, the greater is the real multiplication to which division by $x-2$ is equivalent. If it be said that the quotients x and $x+2$ do not differ by a larger proportion of themselves than x^2-2x and x^2-4, and that, agreeably to the sense preferred in page 24, a is as nearly equal to b, as c is to d, when the differences of the first two and of the last two are in the same proportion as a and c: the answer is, that x^2-2x and x^2-4 must not then be called nearly equal, because they are small, and because their difference is therefore small; for both may be small, and, nevertheless, one may be many times the other. An elephant and a gnat are both small fractions, if the whole earth be called 1, but they are not nearly equal in any sense.

From the above we gather, that, calling a and b nearly equal when they only differ by a small fraction of either, we are not at liberty to say that two small quantities are therefore (because they are small) nearly equal.]

There are two obvious tests of the *absolute* equality of a and b:

$$a - b = 0 \qquad \text{and} \qquad \frac{a}{b} = 1.$$

We lay down the following definition. *Approach towards equality is measured not by the diminution of the difference, but by the approach of the quotient towards* 1. Thus 3 is not more nearly equal to 2 than 20 to 25 because $3 - 2$ is less than $25 - 20$; but 3 is not so nearly equal to 2 as 25 is to 20, because

$$\frac{3}{2} \text{ exceeds 1 by } \frac{1}{2} \qquad \frac{25}{20} \text{ exceeds 1 by } \frac{1}{4} \left(\text{less than } \frac{1}{2}\right)$$

See page 24, for an anticipation of this use of the words "nearly equal," and what precedes in [ ] for reasons.

THEOREM. The value of a fraction depends entirely on the relative, not on the *absolute*, value of the terms. The following examples will shew that this is contained in the theorem $\frac{ma}{mb} = \frac{a}{b}$.

1. Find a fraction whose numerator is 583, and which is as small as $\frac{1}{1000}$. *Answer* $\frac{583}{583000}$.

2. Find two fractions, a and b, each less than $\frac{1}{1000}$, so that $\frac{a}{b}$ may be a million.

Answer $a = \frac{1}{2000}$ $b = \frac{1}{2000,000,000}$.

Here $\frac{a}{b} = 1000,000$ $\qquad \frac{b}{a} = \frac{1}{1000,000}$

3. Find two fractions, a and b, each less than 2, so that $\frac{a}{b}$ may be $= m$.

Answer. Take any two numbers, p and q (only let p be less than $2q$), and let

$$a = \frac{p}{q} \qquad\qquad b = \frac{p}{mq}$$

EXERCISES. (All the letters are positive) p is not so nearly equal to $p + q$ as $p + m$ is to $p + q + m$. If a be more nearly equal to b than c is to e, then $a + c$ is not so nearly equal to $b + e$ as a is to b, but more nearly equal than c is to e. Again, mx is as nearly equal to nx as my is to ny.

We must now say something upon definite and indefinite terms. A word is called *definite* when there can be no question as to whether it is proper or not to use it in any particular case which may be proposed. A word is called *indefinite* when it may be matter of opinion as to whether it is proper or not to use it in any proposed case. Thus, *equal* is a definite term. No two opinions can exist upon the answer to the question "Is $4+4$ *equal* to 9?" But *great* is an indefinite term. "Is 1000 a great number?" The answer to this is, there is no way of answering the question in any case upon which all must agree. We give some examples of each sort.

Definite — equal, exact, larger, nearer, smaller, greater, less, largest, smallest, &c. as large as, as much as, as great as, &c., quite.

Indefinite — near, small, large, great, much, nearly, hardly, &c. large enough, small enough, &c.

With indefinite terms we can have nothing to do, unless by addition to their meaning, so as to give them a signification which will allow us to use them without presenting as mathematical theorems propositions which contain matters of opinion. We shall take the terms *near*, *small*, and *great* as instances. Observe that the term *smaller* is not to be considered as altered in the same manner as *less* in page 62. It keeps its arithmetical meaning. "If x be small, $7+x$ is nearly equal to 7?" This is a proposition in which all will agree: and the reason is that "small" and "near" have a connexion which is independent of what fraction the speaker may choose to think entitled to the term "small." AB may be a line which one may call small, and another not small; but all will agree that in the meaning of the words, "small" and "near" is implied "If AB be *small*, A is *near* to B." But if we come to ask — What fraction is small, is it $\frac{1}{100}$, $\frac{1}{1000}$, &c.? — The answer must depend on circumstances. We reject, therefore, the terms small and near in their common meaning. But the preceding proposition can be put in a form which will never render it necessary to inquire what is small or near. "If x may be as small as we please, then $7+x$ may be made as near as we please to 7;" or, "Let me make x as small as *I* please, and I can make $7+x$ as near to 7 as *you* please; or, "Name any fraction you please, and let it be such a one as you choose to call small, I do not ask why: then, if I may make x as small as I please, I can make $7+x$ differ from 7 by a fraction less than the one named by you;" and so on.

Having rejected the terms *small, great,* and *near*, in their common signification, we shall revive them for our own use in algebra, simply as convenient abbreviations of "as small as we please," "as great as we please," "as near as we please," or, "as small as may be necessary," "as great as" &c. &c. In this sense we have various demonstrable and definite propositions. For instance, "if x be small, $\frac{1}{x}$ is great;" that is, if we may make x as small as we please, we may make $\frac{1}{x}$ as great as we please.

When, under certain circumstances, or by certain suppositions, we can make A as near as we please to P (A being a quantity which changes its value as we alter our suppositions, and P a fixed quantity, which does not change when we alter our suppositions) then P is called the LIMIT *of A.* A quantity which we are supposing as great as we please is said to *increase without limit;* one which we are supposing as small as we please, is said to *decrease without limit.* The following theorems will be evidently true.

If x decrease without limit, the limit of $a+x$ is a; if x diminish without limit, then $\frac{1}{x}$ increases without limit; if x approach without limit towards b, then the limit of $a+x$ is $a+b$.

The first and third may appear but a complicated method of saying that if $x=0$, $a+x=a$; and if $x=b$, $a+x=a+b$, which are perfectly intelligible. But, "if $x=0$ $\frac{1}{x}=\frac{1}{0}$" has no intelligible meaning; in fact, in page 25, we have already anticipated the construction we here put upon that proposition.

One object of this chapter is, to put interpretations upon those forms which would otherwise offer difficulties; such as 0, $\frac{1}{0}$, $\frac{0}{0}$, and (had we not otherwise found a rational interpretation) a^0. But we still have the forms

$$0^0 \qquad 0^{\frac{1}{0}} \qquad \left(\frac{1}{0}\right)^0 \qquad \text{\&c.}$$

all of which might occur, if we stumbled upon such expressions as

$$(x-a)^{x-a} \quad (x-a)^{\frac{1}{x-a}} \quad \left(\frac{1}{x-a}\right)^{x-a} \quad \&c.$$

without observing (what might happen) that $x = a$.

In all these cases, that is, when we get a form which is not a direct representation of quantity, we shall not ask "What is the value of that form?" or in any way enter into the question whether it is demonstrable that it has a value or not. But the question we shall always ask is this: "As we approach the supposition which gives the unintelligible form, to what value does the expression which gives the unintelligible form approach?" For instance,

when $\quad x = a \quad \dfrac{x^2-a^2}{x-a} = \dfrac{0}{0}$

But if we examine what sort of change of value takes place in the above fraction when x approaches towards a, we find that value to approach towards $2a$, as will be afterwards shewn. And it will be found that we have the following proposition: "If x may be made as near as we please to a, then $\dfrac{x^2-a^2}{x-a}$ can be made as near as we please to $2a$," or, "If x approach without limit to a, then $\dfrac{x^2-a^2}{x-a}$ approaches without limit to $2a$."

Shall we then say that

when $x = a \quad \dfrac{x^2-a^2}{x-a} = \dfrac{0}{0} = 2a \quad$ or $\quad \dfrac{0}{0} = 2a$ (in this case)?

Whether it will be proper to say so, in the common meaning of all the terms, we leave to the student. But we shall not, in this work, use such a form, *except as an abbreviation of one of the preceding propositions.*

It is usual to make the symbol ∞ stand for $\dfrac{1}{0}$; and this is called *infinity*. From what has preceded, and page 25, we shall regard $x = \infty$ as an abbreviation of the following: "Let x increase without limit."

Again, "*let* $x = 0$," it will often be most safe to regard as an abbreviation of "let x diminish without limit." We shall hereafter return to this. But in equations of the form $P - Q = 0$, where P and Q are certainly finite quantities, this alteration will not be necessary.

Theorem I. If A and B be two expressions which are always equal, so long as they preserve an intelligible form, then the limits of A and B are also equal.

To take a case, suppose that when x increases without limit, A has the limit P, and B has the limit Q; then P must $=$ Q. To prove this, suppose that

$$A = P + a \qquad\qquad B = Q + b$$

then, by increasing x without limit, a and b are diminished without limit. For, if not, and if a had the limit α, then the limit of A or $P + a$ would be $P + \alpha$. But it is P; therefore a has not a limit, but diminishes without limit.

But because $A = B$ we have $P + a = Q + b$. If P be not $=$ Q it is greater or less. If possible, let P be greater than Q, and let $P = Q + R$. Then, since P and Q are quantities not containing x, R is the same, and P, Q, and R, do not change when x changes. We have then

$$Q + R + a = Q + b \quad \text{or} \quad R = b - a$$

Here, then, is the following absurdity: R (a fixed quantity) is always equal to $b - a$, which can be made as small as we please by increasing x, because b and a diminish without limit as x is increased without limit. Therefore $P = Q + R$ is absurd. In a similar way it may be proved that $P = Q - R$ is absurd. Therefore $P = Q$.

Theorem II. When x diminishes without limit, the way of finding the limit of an expression is to make $x = 0$, provided, 1st, all the results be intelligible; 2d, that the number of operations be not unlimited.

For instance, it is clear enough that $1 + 2x + 3x^2$, when x diminishes without limit, has the limit $1 + 0 + 0$ or 1. And, perhaps, the student may think it clear that the limit of

$$1 + x + x^2 + x^3 + x^4 + \text{ \&c. continued for ever}$$

is

$$1 + 0 + 0 + 0 + 0 + \text{ \&c. } \ldots\ldots\ldots\ldots\ldots$$

or 1, when x diminishes without limit. But here we must make him observe, that when we take x small, though each of the terms x, x^2, x^3, x^4, &c., may be small, yet their number is unlimited. And though we know that when a *certain number* of terms is added together, each of which may be made as small as we please, that their sum can be made as small as we please, yet we do not know the same of an *unlimited number* of terms.

Theorem III. When x increases without limit, it is clear that such expressions as

$$a + bx, \quad a + bx + cx^2, \text{ \&c. increase without limit:}$$

and that $a + \frac{b}{x}$, $a + \frac{b}{x} + \frac{c}{x^2}$, &c. have the limit a.

But the easiest way, in general, to examine expressions, will be to remember that when x increases without limit, $\frac{1}{x}$ decreases without limit.

Let, then, $v = \frac{1}{x}$ or $x = \frac{1}{v}$: substitute this value of x, and, if possible, reduce the expression to a form in which its limit will be evident when v diminishes without limit, that is, when x increases without limit.

For instance, what is the limit of $\frac{x+1}{3x-2}$ in such a case.

$$\text{Let } x = \frac{1}{v} \qquad \frac{\frac{1}{v}+1}{\frac{3}{v}-2} = \frac{\left(\frac{1}{v}+1\right)v}{\left(\frac{3}{v}-2\right)v} = \frac{1+v}{3-2v}$$

When v diminishes without limit, the preceding has the limit $\frac{1}{3}$.

Let the student now prove the following cases, which we express in the abbreviated form.

$$\text{If} \quad x = \infty \quad \frac{ax^2 + bx + c}{px + q} = \infty \quad \frac{ax^2 + bx + c}{px^2 + qx + r} = \frac{a}{p}$$

$$\frac{ax^2 + bx + c}{px^3 + qx + r} = 0$$

THEOREM IV. If a be greater than 1, the terms of the series, a, a^2, a^3, a^4, &c., increase without limit; or (abbreviated) $a^\infty = \infty$

For	$a^2 = a + a^2 - a = a + a(a-1)$	
or	a becomes a^2	by adding $a\ (a-1)$
Similarly	a^2 becomes a^3	by adding $a^2\ (a-1)$
	....	
Generally	a^n becomes a^{n+1}	by adding $a^n\ (a-1)$

But because a is greater than 1, $a-1$ is positive, and the addition of $a(a-1)$ is therefore arithmetical increase. And a^2 is greater than a; therefore $a^2(a-1)$ is greater than $a(a-1)$, or the third power of a exceeds the second by more than the second exceeds the first. Similarly, the fourth exceeds the third by more than the third exceeds the second; and so on. But if to a the same quantity be added as many times as we please, the result may be made as great as we please; still more if the same number of additions be made, with a greater quantity each time than the last. Whence follows the theorem.

THEOREM V. If b be less than 1, the terms of the series b, b^2, b^3, b^4, &c. decrease without limit, or (abbreviated) $b^{\infty} = 0$.

Let $b = \frac{1}{a}$, then $b^n = \frac{1}{a^n}$. But because b is less than 1, a or $\frac{1}{b}$ is greater than 1; therefore a^n may be made (Theorem IV.) as great as we please. Hence, $\frac{1}{a^n}$ or b^n may be made as small as we please.

Find out the first of the powers of $\frac{9}{100}$ which is less than $\frac{1}{1000000}$.

Ans. The sixth.

It is hardly necessary to notice, that if $a = 1$ the terms of the series a, a^2, a^3, &c. neither increase nor decrease.

THEOREM VI. If x be positive, and less than 1, the series of terms

$$(1+x) \quad (1+x+x^2) \quad (1+x+x^2+x^3) \text{ \&c.}$$

increases, but not without limit. The limit is $\frac{1}{1-x}$; that is, no term of the preceding, how many powers soever it may contain, can be as great as $\frac{1}{1-x}$, but may come as near to it as we please. The abbreviation is as follows:

$$\frac{1}{1-x} = 1 + x + x^2 + x^3 + \ \ldots\ + x^{\infty}$$

more generally written

$$\frac{1}{1-x} = 1 + x + x^2 + x^3 + \text{ \&c. } \textit{ad infinitum.}$$

As this series is a most important part of the groundwork of all that follows, we shall try to establish the proposition from the method of its formation. We remark that when x is positive, the terms 1, $1 + x$, $1 + x + x^2$, &c. evidently increase; and that each term is formed by multiplying the preceding by x, and then adding 1. Thus, $1+x+x^2$ is $1+x(1+x)$, and $1+x+x^2+x^3$ is $1+x(1+x+x^2)$, and so on. If A stand for any term, and B for the next; then

$$B = 1 + Ax$$

Now, B is greater than A; therefore, adding 1 more than compensates the diminution which A undergoes by being multiplied by x

(Remember that x is less than 1). But $Ax = A + Ax - A = A - (1-x)A$; or multiplication into x diminishes A by $(1-x)$A. This the addition of 1 more than compensates; that is, 1 is greater than $(1-x)$A. Divide both of these by $1-x$; and $\frac{1}{1-x}$ is greater than A. But A is any term we please of 1, $1+x$, $1+x+x^2$, &c.; therefore every one of these, how far soever we go, is less than $\frac{1}{1-x}$.

Now we have to prove that, though we cannot find A as great as $\frac{1}{1-x}$, we can come as near to this as we please. Remember that the way of forming the next term is always, "Multiply by x and add 1." Let A differ from $\frac{1}{1-x}$ by p; so that

$A = \frac{1}{1-x} - p$, the next term is $1 + Ax = 1 + \frac{x}{1-x} - px$

$= \frac{1-x+x}{1-x} - px = \frac{1}{1-x} - px$; the next term is

$1 + \frac{x}{1-x} - px^2$ or $\frac{1}{1-x} - px^2$; the next term is

$\frac{1}{1-x} - px^3$; and so on.

Hence we can find a term which differs from $\frac{1}{1-x}$ by px^n, where n is as great as we please. But p is a given quantity, and x^n (Theorem V.) diminishes without limit when n increases without limit; therefore px^n may be made as small as we please, or $\frac{1}{1-x} - px^n$ as near to $\frac{1}{1-x}$ as we please. But, by continuing the preceding terms from A, we shall at last come to $\frac{1}{1-x} - px^n$. Therefore, by continuing the terms, we come as near to $\frac{1}{1-x}$ as we please.

We shall now try to find whether (x being less than 1),

$$1 - x + x^2 - x^3 + x^4 - \text{ \&c. continued } \textit{ad infinitum},$$

has a limit; or what is the nature of the increase or decrease of 1, $1-x$, $1-x+x^2$, &c. Here we see alternate increase and decrease: but still under a simple law. To find the next term, multiply the last term by x, and take the result from 1. Thus,

$$1 - x + x^2 = 1 - x(1 - x)$$

$$1 - x + x^2 - x^3 = 1 - x(1 - x + x^2) \text{ and so on.}$$

Or if A and B be two consecutive terms,

$$B = 1 - Ax \quad \text{or} \quad 1 + A - A(1 + x)$$

That is $$B = A + 1 - A(1 + x)$$

then the next term $C = B + 1 - B(1 + x)$ &c. &c.

But the results alternately increase and decrease; that is, suppose B greater than A, then C is less than B. Or, suppose 1 greater than $A(1 + x)$, then 1 is less than $B(1 + x)$: or $\frac{1}{1+x}$ is greater than A, and $\frac{1}{1+x}$ is less than B. So that the results are alternately less and greater than $\frac{1}{1+x}$.

Thus we have

1	is greater than	$\frac{1}{1+x}$
$1 - x$	is less than	$\frac{1}{1+x}$
$1 - x + x^2$	is greater than	$\frac{1}{1+x}$
&c.	&c.	&c.

Now, since x^n diminishes without limit as n is increased, we can take n so great that two consecutive terms

$$1 - x + x^2 - \text{\&c.} \cdots\cdots \pm x^{n-1}$$

and $$1 - x + x^2 - \text{\&c.} \cdots\cdots \pm x^{n-1} \mp x^n$$

shall differ by a quantity as small as we please (for they differ by $\mp x^n$). But we have just proved that one of these terms is greater than $\frac{1}{1+x}$ and the other less. And they differ less from any quantity which falls between them, than they do from each other; consequently, either may be (if n be taken sufficiently great) as near to $\frac{1}{1+x}$ as we please.

These two results we express as follows:

$$\frac{1}{1-x} = 1 + x + x^2 + x^3 + \text{\&c. } \textit{ad infinitum.}$$

$$\frac{1}{1+x} = 1 - x + x^2 - x^3 + \text{\&c. } \textit{ad infinitum.}$$

and $\frac{1}{1-x}$ is called the *sum* of the first infinite series, meaning, the limit towards which we may come as near as we please by continual addition of the terms $1, x, x^2, x^3$, &c.

We shall proceed with this subject in chapter VIII.

THEOREM VII. If the numerator and denominator of a fraction diminish without limit, the limit of that fraction may be nothing, finite or infinite: that is, the fraction may diminish without limit, may have a finite limit, or may increase without limit. Neither of these suppositions is inconsistent with the unlimited diminution of both the numerator and denominator.

Take the three fractions

$$\frac{x^2-a^2}{(x-a)^2} \qquad \frac{x^2-a^2}{x-a} \qquad \frac{(x-a)^2}{x^2-a^2}$$

By supposing $x = a$, all the three assume the form $\frac{0}{0}$. By supposing x to approach as near as may be necessary to a, we may diminish the numerators and denominators without limit. The reason of this is, that $x-a$ is a factor of every numerator and every denominator, and $x-a$ diminishes without limit as x approaches to a. For the three fractions are

$$\frac{(x-a)(x+a)}{(x-a)(x-a)} \qquad \frac{(x-a)(x+a)}{(x-a)} \qquad \frac{(x-a)(x-a)}{(x-a)(x+a)}$$

Divide both terms of each fraction by $x-a$, which gives

$$\frac{x+a}{x-a} \qquad x+a \qquad \frac{x-a}{x+a}$$

Which are always severally equal to the first, except where $x = a$, on which we give no opinion (see page 156). But as x approaches towards a, the first is

$$\frac{\text{A quantity whose limit is } 2a}{\text{A quantity which diminishes without limit,}}$$

and therefore increases without limit. The second is

A quantity whose limit is $2a$,

and therefore approaches without limit to $2a$. The third is

$$\frac{\text{A quantity which diminishes without limit}}{\text{A quantity whose limit is } 2a,}$$

and therefore diminishes without limit. Consequently, when we see a fraction under circumstances in which the numerator and denominator diminish without limit, we have no right to draw any conclusion as to the value towards which that fraction is tending, but must examine the fraction itself to see whether it diminishes or increases without limit, or whether it tends towards a finite limit.

THEOREM VIII. The same caution is necessary as to a fraction whose terms increase without limit, or which approaches the form $\frac{\infty}{\infty}$. For let $\frac{A}{B}$ be a fraction whose terms increase without limit. We know that

$$\frac{A}{B} = \frac{\frac{1}{B}}{\frac{1}{A}}$$

and when A and B increase without limit, $\frac{1}{A}$ and $\frac{1}{B}$ diminish without limit. Therefore the same circumstances which make $\frac{A}{B}$ approach the form $\frac{\infty}{\infty}$, make the same fraction (in a different form) approach to $\frac{0}{0}$. Whence the last theorem applies.

THEOREM IX. The same caution applies to the value of a product, in which one of the terms diminishes without limit, while the other increases without limit. Let AB be such a product, which approaches the form $0 \times \infty$: that is, while A diminishes without limit, B increases without limit. We know that

$$AB = \frac{A}{\frac{1}{B}}$$

and when B increases without limit, $\frac{1}{B}$ diminishes without limit. Hence, as in the last case, A B, under a different form, approaches to the form $\frac{0}{0}$.

Thus we see that the three forms

$$\frac{0}{0} \qquad \frac{\infty}{\infty} \qquad 0 \times \infty$$

are so connected, that any expression which gives one, may be made to give either of the others.

We now take the form a^0, considered, not in the absolute and defined sense of page 85, but as the representative of

The limit of a^x when x diminishes without limit.

If $x = \frac{1}{y}$, then when x diminishes without limit, y increases without limit. Let y be a whole number, then

$$a^x = a^{\frac{1}{y}} = \sqrt[y]{a}$$

Firstly, if a be greater than 1, all its roots are greater than 1 (for all the powers of less than 1 are less than 1). Let

$$\sqrt[y]{a} = 1 + v \qquad \text{or} \qquad a = (1 + v)^y$$

Now, y can be taken so great that v shall be less than any fraction we may name, however small. For, if not, say it is impossible that y should be so great as to make v less than k. Then, v being always greater than k, whatever may be the value of y, $1 + v$ is always greater than $1 + k$. But, Theorem IV., y may be taken so great that $(1+k)^y$ shall exceed any quantity we may name, and shall therefore exceed a. Still more, then, will $(1+v)^y$ exceed a (v is greater than k). But $(1+v)^y$ *equals* a: here then is a contradiction. Consequently, the supposition that v can never be made less than a given fraction k is not true: that is, v can be made less than any given fraction, or $1 + v$ can be brought as near to 1 as we please. But $1 + v = \sqrt[y]{a}$, therefore, if y increase without limit, we see that

$$\sqrt[y]{a} \text{ or } a^{\frac{1}{y}} \text{ or } a^x \text{ has the limit } 1$$

Or $a^0 = 1$ where 0 is used in the sense given in page 156.

Secondly, let a be less that 1, whence $\frac{1}{a}$ is greater than 1. By the last case $\sqrt[y]{\frac{1}{a}}$ can be brought as near to 1 as we please; but this is $1 \div \sqrt[y]{a}$, therefore $\sqrt[y]{a}$ can be brought as near to 1 as we please.

Theorem X. In any rational* integral expression with respect to x, if x may be increased without limit, the term which has the highest power of x may be made to contain the sum of all the rest times without limit. That is, in the expression

$$ax^3 + bx^2 + cx + e, \quad \text{for example,}$$

let a be any given quantity, however small, and b, c, and e, any given

* Look at the beginning of the next chapter.

quantities, however great, yet x may be taken so great that ax^3 shall contain bx^2+cx+e as many times as we please.

The number of times and parts of a time which ax^3 contains bx^2+cx+e is expressed by the fraction

$$\frac{ax^3}{bx^2+cx+e} \text{ or } \frac{ax^3 \div x^2}{(bx^2+cx+e)\div x^2} \text{ or } \frac{ax}{b+\frac{c}{x}+\frac{e}{x^2}}$$

Let $\frac{c}{x}+\frac{e}{x^2}=p$. Then, by increasing x without limit, p is diminished without limit, and will, if x be taken sufficiently great, become less than 1. That is, ax^3 contains bx^2+cx+e, $\frac{ax}{b+p}$ times, or more than $\frac{ax}{b+1}$ times. But $\frac{ax}{b+1}$ or $\frac{a}{b+1}\times x$ increases without limit, when x increases without limit. The same then does the number of times which ax^3 contains bx^2+cx+e.

For example, how great must x be, in order that we may be certain that the millionth part of x^3 contains $1000x^2+500x+1000$ more than a hundred thousand times?

$$\frac{\text{one millionth of } x^3}{1000x^2+500x+1000} = \frac{\text{one millionth of } x}{1000+\frac{500}{x}+\frac{1000}{x^2}}$$

Now, if x be 1000 or more, $\frac{500}{x}+\frac{1000}{x^2}$ is less than 1. Therefore, in this case, the preceding fraction is greater than one millionth of $x \div 1001$, or than

$$\frac{x}{1001000,000}$$

If we take $x=1001000,000\times 100,000$ or $100100,000,000,000$, the preceding fraction becomes 100,000. Hence, the millionth part of x^3 is greater than 100,000 times $(1000x^2+500x+1000)$. We do not say that this is the least value of x which will answer the conditions, but that this, or any thing greater, will do so.

Theorem XI. In any integral and rational expression with respect to x, if x may be diminished without limit, the term containing the lowest power of x may be made to contain the rest of the expression as many times as we please. For instance, in $\frac{1}{1000}x+1000x^2+100x^3$ we may take x so small that $\frac{1}{1000}x$ shall contain $1000x^2+100x^3$ as many times as we please: or in

ax^3+bx^2+cx+e, x may be taken so small that e (or ex^0, which is the term containing the lowest power* of x) shall contain ax^3+bx^2+cx as often as we please. This last is evident in this particular case, because e remains the same when x diminishes without limit, and ax^3+bx^2+cx diminishes without limit. Therefore, the second may become less than any given fraction of the first. Now, take a case in which there is no power of x so low as x^0; such as ax^3+bx^2+cx. Here x may be taken so small that cx shall contain ax^3+bx^2 as many times as we please. For the number of times and parts of times which cx contains ax^3+bx^2 is

$$\frac{cx}{ax^3+bx^2} \text{ or } \frac{c}{ax^2+bx} = \frac{\text{fixed quantity}}{\text{one which diminishes without limit}}$$

which latter increases without limit when x is diminished without limit.

Hence it follows, that when x increases without limit, x^2, x^3, x^4, &c. not only increase without limit, but *each of them increases without limit with respect to the preceding;* by which is meant that x^3 increases so much faster than x^2, that x^3 will come at last to contain x^2 as many times as we please. Similarly, when x diminishes without limit, we find that x^2, x^3, x^4, &c. not only diminish without limit, but *each of them diminishes without limit with respect to the preceding;* by which is meant that x^3 diminishes so much faster than x^2, that x^3 will come at last to be as small a fraction of x^2 as we please. These notions are sometimes abbreviated into the following phrases, which it must be remembered are not intelligible, *except as abbreviations.*

Abbreviated Phrases.	Rational Meaning.
1. Of two infinitely great quantities, one may be infinitely greater than the other.	1. Of two quantities which increase without limit, one may increase so much faster than the other, as not only to increase without limit absolutely speaking, but to increase without limit in the number of times which it contains the other.

* The algebraical series of whole powers of x is

$$\ldots\ldots \quad x^{-3} \quad x^{-2} \quad x^{-1} \quad x^0 \quad x^1 \quad x^2 \quad x^3 \quad \ldots\ldots$$

answering to

$$\frac{1}{x^3} \quad \frac{1}{x^2} \quad \frac{1}{x} \quad 1 \quad x \quad x^2 \quad x^3 \quad \ldots\ldots$$ (see page 85.)

Abbreviated Phrases.	Rational Meaning.
2. Of two infinitely small quantities, one may be infinitely less than the other.	2. Of two quantities which diminish without limit, one may diminish so much faster than the other, as not only to diminish without limit absolutely speaking, but to diminish without limit in the fraction which it is of the other.

Let the student now try if he can explain the following Problem.

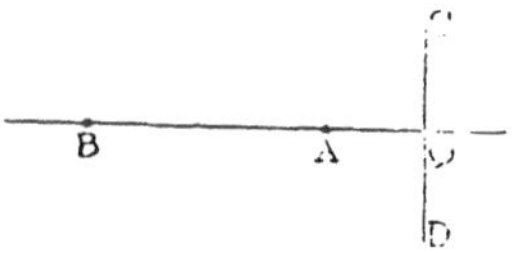

If A and B move together towards the line CD, and if A move in such a way that AQ is $\frac{1}{4}$ of an inch when BQ is $\frac{1}{2}$ an inch; that AQ is $\frac{1}{9}$ of an inch when BQ is $\frac{1}{3}$ of an inch; or generally that AQ is x^2 inches* when BQ is x inches: and if a microscope be placed over the figure which grows in magnifying power as A moves towards Q, in such a way that the increase of magnifying power just compensates the real diminution of AQ, so that AQ always appears of the same length; then B, instead of appearing to move toward Q, will appear to move away from Q.

* It is usual to say x *inches*, when x is less than 1, or when x inches is really a fraction of an inch; in which case x *of an inch* would be more agreeable to analogy. Returning to the consideration discussed in the note to page 40, it will be useful to observe that the idiom of our language makes the connexion between multiplication by 4 and multiplication by $\frac{1}{4}$, less obvious than it might have been. We only say 4 *of* before an article or pronoun, "four of the men," "four of them." But we always say, "one fourth *of* six," "one fourth *of* an inch." If it had been idiomatic to say "four of six," or "a four of sixes," for 24; and "four *of* an inch" for "four inches," the propriety of extending the term multiplication to fractions would have been much more obvious.

CHAPTER VII.

CLASSIFICATION OF ALGEBRAICAL EXPRESSIONS AND CONSEQUENCES. RULE OF DIVISION.

Previously to commencing this subject, we shall make a classification of the expressions we have obtained. All the terms in use are usually relative to some particular letter; in what follows we shall suppose this letter to be x. We shall proceed to explain the following table.

Functions.

- Common Algebraical.
  - Rational.
    - Integral.
      - Monomial.
      - Binomial,
      - Trinomial,
      - Quadrinomial,
      - &c.
    - Fractional.
  - Irrational.
    - Integral.
      - Monomial,
      - Binomial,
      - Trinomial,
      - Quadrinomial,
      - &c.
    - Fractional.
- Transcendental.
  - Exponential,
  - Logarithmic,
  - Circular,
  - Inverse Circular,
  - &c.

Any expression which contains x in any way is called a *function* of x: thus $a+x$, $a+bx^2$, &c. are functions of x; they are also functions of a and b, but may be considered only with regard to x. All expressions which contain only a *finite* number of such expressions as are treated in the preceding part of this work, are called *common algebraic** functions, except only where x is an exponent. Thus $\sqrt{a+x^2}$, ax^3+b, &c. are common algebraic functions, but a^x is not. And we do not know whether $1+x+x^2+$ &c. *ad infinitum* (page 159) is a common algebraic function of x or not, until we have found that it is the same as $1\div(1-x)$. All other functions of x are called *transcendental* functions; such are a^x and all functions containing it; such will be (when we come to define them) the

* Usually, *algebraic* functions; but transcendental functions are certainly also *algebraic*, that is, considered in algebra.

logarithm of x, the sine and cosine of x in trigonometry, and many others. A function which contains a^x is called an *exponential** function of x, one which contains a logarithm, a logarithmic function, &c.

Common algebraical functions are divided into *rational*, which contain only whole powers of x, as $a+x^2$, $ax^{-2}+b$, &c.; and *irrational*, which contain roots or fractional powers of x, as $ax^{\frac{1}{2}}+b$, $\sqrt{a^2+x^2}$, &c.

Rational and irrational functions of x are both divided into *integral*, which contain x only in numerators, as

$$x+\frac{x^2}{a} \quad \text{and} \quad \frac{a+\sqrt{x}}{a+b};$$

and *fractional*, which contain x in denominators,

$$\text{as} \quad \frac{a+x}{bx+x^2} \quad \text{and} \quad \frac{\sqrt{x}-\sqrt{y}}{c+\sqrt[3]{x}}$$

Integral functions are divided into *monomials*, which contain only one power of x, as x^3, ax^2, $\sqrt{bx}$, $(a+b)x^4$; *binomials*, which contain two distinct powers of x (x^0 included) as $a+bx$ or ax^0+bx, $cx^3+\sqrt{x}$, mx^2+nx^3, &c.; *trinomials*, containing three distinct powers; *quadrinomials* containing four, &c. The two latter terms are little used: all expressions of more than one term are called *polynomials*.

Integral and rational functions are divided into those of the first, second, third, &c. degrees, according to the exponent of the highest power which is found in them. Thus

$a+bx$ is a rational integral function of x, of the first degree.

$a+bx+cx^2$ .... of the second degree.

&c. &c.

The term a, if written ax^0, is of *no* degree with respect to x.

The expression† $\dfrac{a+\sqrt{b}\log c+a^2c^x}{m+\sqrt{n}}$

* The letter principally considered is an exponent.

† The meaning of log. c, or logarithm of c, will be afterwards explained.

is a rational integral trinomial function of	a
.... rational fractional	m
.... irrational integral	b
.... irrational fractional	n
.... exponential	x
.... logarithmic	c

Rational and integral functions are generally arranged, so that the powers of x may rise or fall continually in going from left to right. Thus, $ax+b-cx^2$ is never so written, but either

$$-cx^2+ax+b \quad \text{or} \quad b+ax-cx^2$$

in the first case it is said to be *arranged in descending*, in the second in *ascending*, powers of x.

Thus $\qquad a-bx^3+cx-x^5-x^3$

should be $\qquad a+cx-(b+1)x^3-x^5$

or $\qquad -x^5-(b+1)x^3+cx+a$

The most important class of functions is the rational and integral, containing those which appear rational and integral, but of which it cannot be known whether they are algebraical or transcendental, owing to their containing an infinite number of terms. Such are the forms

$$a+bx+cx^2+ \quad \cdots\cdots\cdots \quad +px^{n-1}+qx^n$$

where a, b, c are not functions of x, and n is a whole number; and

$$a+bx+cx^2+ex^3+ \quad \cdots\cdots \quad + \text{\&c. } \textit{ad infinitum}.$$

The reduction of expressions to such forms is one of the principal branches of the subject. We shall call the first generally a *polynomial*, the second an *infinite series*.

DEFINITIONS. In multiplication of polynomials, the several products formed in the process may be called *subordinate products*. Thus, in multiplying $a+x$ by $b+x$, the subordinate products are ab, ax, bx, and x^2. A *term* of a polynomial is all that contains any one power of x; thus, the preceding product $ab+ax+bx+x^2$ is not said to be of four terms, but of three, namely, ab, $(a+b)x$, and x^2.

THEOREM. In the product of two polynomials, there must be at least two terms, which are subordinate products, and not formed by two or more subordinate products.

Suppose we multiply $ax+bx^2+cx^3$ and px^4+qx^5. It is plain that no other subordinate product can contain so high a power of x as $cx^3 \times qx^5$, or cqx^8, nor so low a power of x as $ax \times px^4$, or apx^5, because, in these, both exponents are the highest in their expressions, or the lowest. Consequently, these must be terms of the product, which is, in fact,

$$apx^5 + (aq+bp)x^6 + (bq+cp)x^7 + cqx^8$$

having four terms, two of which are the simple subordinate products already noticed.

Algebraical division differs from arithmetical in this, that in the latter we wish to ascertain whether a whole number P can be made by taking another whole number Q a whole number of times; in the former we inquire whether a polynomial function of x, P, can be made by multiplying another polynomial Q by any third polynomial. For instance, to divide $8x^3+1$ by $2x+1$, is the following: To reduce $\frac{8x^3+1}{2x+1}$ if possible, to a simple polynomial. The way of treating this question will also give that of treating any other.

If possible, let $8x^3+1$ be made by multiplying $2x+1$ by $a+bx+cx^2+ex^3+$ &c. Now, firstly, this latter expression cannot go higher than cx^2; for, if it did, say to ex^3, we should have, by the last theorem, $ex^3 \times 2x$ or $2ex^4$ in the product. But that product is $8x^3+1$, in which x^4 does not appear; consequently, ex^3 and higher terms are not in the polynomial required, which is therefore of the form $a+bx+cx^2$. We have then, if our question be possible,

$$8x^3+1 = (2x+1)(cx^2+bx+a)$$

We have proved that $2x \times cx^2$ must be a term* of this product; but it can only be $8x^3$, therefore $2x \times cx^2 = 8x^3$, or $cx^2 = 8x^3 \div 2x = 4x^2$.

Consequently, $$8x^3+1 = (2x+1)(4x^2+bx+a)$$
$$= (2x+1)4x^2 + (2x+1)(bx+a)$$

$$8x^3+1-(2x+1)4x^2 \quad \text{or} \quad -4x^2+1 = (2x+1)(bx+a)$$

of the latter product, $2x \times bx$ must be a term; but it can only be $-4x^2$, therefore $bx = -4x^2 \div 2x = -2x$, or

$$-4x^2+1 = (2x+1)(-2x+a)$$
$$= -2x(2x+1) + (2x+1)a$$

$$-4x^2+1+2x(2x+1) \quad \text{or} \quad (2x+1) = (2x+1)a$$

* This being a subordinate product, which cannot be altered or destroyed by any other subordinate product.

This last equation is made identical by $a=1$; therefore $4x^2-2x+1$ is the polynomial by which $2x+1$ being multiplied, the product is $8x^3+1$: as will be found by trial.

The steps of the preceding process may be arranged after the manner of division in arithmetic; the only difference being, that instead of finding a new term in the quotient by trial of the left hand figures of the divisor and dividend, it is found by dividing the left hand term of the dividend or remainder by that in the divisor. As follows, in which the same question is solved in the two different arrangements.

$$\begin{array}{rl} 2x+1)8x^3+1(4x^2-2x+1 & \quad 1+2x)1+8x^3(1-2x+4x^2 \\ 8x^3+4x^2 & \quad 1+2x \\ \hline -4x^2+1 & \quad -2x+8x^3 \\ -4x^2-2x & \quad -2x-4x^2 \\ \hline 2x+1 & \quad 4x^2+8x^3 \\ 2x+1 & \quad 4x^2+8x^3 \\ \hline 0 & \quad 0 \end{array}$$

Great care must be taken to preserve the same order of arrangement throughout, either in ascending or descending powers of x. The following is the general theory of this process:

First, it is evident that the sum, difference, and product of rational polynomials, are rational polynomials. Let P and Q be two rational polynomials, from which it is desired to obtain V in such a way that $P = QV$. Here P is the dividend, Q the divisor, and V the quotient to be found. Assume any convenient polynomial or monomial A, multiply it by Q, and subtract the product from P, which gives $P-AQ$. Call this R, so that

$$P-AQ = R \quad \dots\dots\dots\dots \quad (1)$$

Assume any other polynomial A′; repeat the process with R (instead of P) and Q. Call the result R′:

$$R-A'Q = R' \quad \dots\dots\dots\dots \quad (2)$$

Assume a third polynomial A″, and let

$$R'-A''Q = R'' \quad \dots\dots\dots\dots \quad (3)$$

The object is to simplify the remainder at every step, so as to reduce it at last to a form in which we may see one of these two things; either how to find a new polynomial which shall reduce the next remainder to 0, or that this is impossible. Suppose, first, that a new polynomial or monomial A‴ can be found, which will reduce the remainder to 0.

$$R'' - A'''Q = 0 \;\ldots\ldots\ldots\; (4)$$

We have, then, from the different equations

$$P = AQ + R = AQ + A'Q + R' = AQ + A'Q + A''Q + R''$$
$$= AQ + A'Q + A''Q + A'''Q = Q(A + A' + A'' + A''')$$

so that $A + A' + A'' + A'''$ is the polynomial required. Suppose that instead of (4), we have

$$R'' - A'''Q = R'''$$

and suppose it to be evidently useless to attempt to continue the process further. We have then

$$P = AQ + A'Q + A''Q + A'''Q + R'''$$

$$(\div)Q \quad \frac{P}{Q} = A + A' + A'' + A''' + \frac{R'''}{Q}$$

$$= \text{rational polynomial} + \left\{ \begin{array}{l} \text{a more simple} \\ \text{fraction than } \frac{P}{Q} \end{array} \right.$$

EXAMPLE. To reduce $\dfrac{x^5+1}{x^2+2x}$ to a more simple form

$$P = x^5 + 1 \qquad\qquad Q = x^2 + 2x$$

$$x^2 + 2x)x^5 + 1(A = \frac{x^5}{x^2} = x^3$$
$$x^5 + 2x^4 = AQ$$
$$-2x^4 + 1 = R, \quad A' = \frac{-2x^4}{x^2} = -2x^2$$
$$-2x^4 - 4x^3 = A'Q$$
$$4x^3 + 1 = R', \quad A'' = \frac{4x^3}{x^2} = 4x$$
$$4x^3 + 8x^2 = A''Q$$
$$-8x^2 + 1 = R'', \quad A''' = -\frac{8x^2}{x^2} = -8$$
$$-8x^2 - 16x = A'''Q$$
$$16x + 1 = R'''$$

It is useless to carry the process further, and we have

$$\frac{x^5+1}{x^2+2x} = x^3-2x^2+4x-8+\frac{16x+1}{x^2+2x}$$

From the preceding, we may deduce the following theorem, which is useful in many parts of mathematics.

If P and Q be two rational polynomials of which P is of the higher degree (or *dimension*, as it is frequently called), then $\frac{P}{Q}$ can be reduced to the form $G+\frac{H}{Q}$ where G and H are rational polynomials, and H of at least one dimension less than Q.

EXERCISE. If, in the preceding process, the remainders be severally multiplied by B, B′, B″, &c. before using them, then

$$\frac{P}{Q} = A+\frac{A'}{B}+\frac{A''}{BB'}+\frac{A'''}{BB'B''}+\frac{R'''}{BB'B''Q}$$

The preceding process may be used in an infinite number of different ways; for though it is only convenient to employ it as in the preceding example, yet in the reasoning P, Q, A, A′, &c. may be any quantities whatsoever. As in the following example,

$$P = 1 \qquad\qquad Q = 1+x$$

$$\begin{array}{l} 1+x)\ 1\ \ (\text{let } A = x \\ \quad\quad\ \ \underline{x+x^2} \\ \quad\quad\ \ 1-x-x^2 = R, \quad \text{let } A' = x^2 \\ \quad\quad\ \ \underline{x^2+x^3} \\ \quad\quad\ \ 1-x-2x^2-x^3 \end{array}$$

$$\frac{P}{Q} = \frac{1}{1+x} = x+x^2+\frac{1-x-2x^2-x^3}{1+x}$$

But this would in most cases amount to no more than an arbitrary method of adding or substracting fractions. When, however, the process of dividing the left-hand term of the remainder by that of the divisor is followed, the result will generally be a symmetrical, and often a useful, developement. For instance, we thus obtain

$$\frac{1}{1+x} = 1-x+x^2-x^3+x^4-\frac{x^5}{1+x}$$

$$\frac{1}{1-x} = 1+x+x^2+x^3+\frac{x^4}{1-x}$$

$$\frac{1}{x+1} = \frac{1}{x} - \frac{1}{x^2} + \frac{1}{x^3} - \frac{1}{x^4} + \frac{1}{x^4}.\frac{1}{x+1}$$

$$\frac{1}{1-2x+x^2} = 1+2x+3x^2+4x^3+\frac{5x^4-4x^5}{1-2x+x^2}$$

By this method we can tell whether either of the preceding series of terms continued without limit will approach a limit or not. For instance, we see that

$$1+x+x^2+x^3+\cdots\cdots+x^n\cdots\cdots(\text{A})$$

becomes $\frac{1}{1-x}$ when $\frac{x^{n+1}}{1-x}$ is added to it. If then x be less than 1, since x^{n+1} diminishes without limit (page 159) when n is increased without limit, the sum of the terms in (A) continually approaches to $\frac{1}{1-x}$ as was shewn in page 160.

Let us, *for the present*, denote by (P) that P is a rational polynomial; and by $(\text{P})+(\text{Q}) = (\text{P}+\text{Q})$, that P and Q are rational polynomials, whence their sum is a rational polynomial. We have then, *always*,

$$(\text{P})+(\text{Q}) = (\text{P}+\text{Q}), \quad (\text{P})-(\text{Q}) = (\text{P}-\text{Q}),$$

$$(\text{P})\times(\text{Q}) = (\text{PQ}); \quad \textit{and} \quad \frac{(\text{P})}{(\text{Q})} = \left(\frac{\text{P}}{\text{Q}}\right) \textit{ in certain cases.}$$

Every polynomial which divides (P) without remainder is called a *factor* of P; thus $x^2-1 = (x+1)(x-1)$ and $x+1$ and $x-1$ are factors of x^2-1. It is evident, from page 171,

1. That no polynomial can have a factor of a higher dimension than its own dimension.

2. That if the polynomial be of m dimensions, and one of its two factors of p dimensions, the remaining factor must be of $m-p$ dimensions.

Thus in

$$\begin{aligned} x^4-1 &= (x-1)(x^3+x^2+x+1) \\ &= (x^2-1)(x^2+1) \end{aligned}$$

the polynomial being of the fourth degree, its factors are in one case of the *first* and *third* $(1+3=4)$, and in the other of the *second* and *second* $(2+2=4)$.

In what follows we speak only of rational polynomials, and by *rational* division we mean that the preceding process leaves no remainder. The following theorems are evident consequences of what goes before.

1. Rational division is impossible unless the dividend be at least as high in degree as the divisor.

2. Where the dividend is higher than the divisor, and rational division is still impossible, the remainder is of a lower degree than either the divisor or dividend.

3. If the dividend be of the mth and the divisor of the nth degree, the quotient is of the $(m-n)$th, and the remainder not higher than of the $(n-1)$th degree. (For so long as the remainder is as high or higher than the divisor, the process can be continued).

4. The dividend being P, the divisor Q, the quotient* A, and the remainder R, then

$$P = AQ+R \quad \text{or} \quad \frac{P}{Q} = A+\frac{R}{Q}$$

5. Every quantity which rationally divides M and N rationally divides their sum, difference, and product. Let Z be the divisor; then

$$\frac{M}{Z} = (A) \quad \frac{N}{Z} = (B) \qquad \frac{M+N}{Z} = (A+B)$$

$$\frac{M-N}{Z} = (A-B) \quad \frac{MN}{Z} = ABZ = (ABZ)$$

6. Every divisor of P and Q (in 4.) divides R, and every divisor of Q and R divides P, &c. so that no two of the three has any rational divisor which the third has not. For instance, let Z divide P and Q rationally, then

$$\frac{P}{Z} \text{ is rational or } = \left(\frac{P}{Z}\right) \qquad \frac{Q}{Z} = \left(\frac{Q}{Z}\right)$$

$$A\times\left(\frac{Q}{Z}\right) = \left(\frac{AQ}{Z}\right) \text{ therefore } \left(\frac{P}{Z}\right)-\left(\frac{AQ}{Z}\right) = \left(\frac{P}{Z}-\frac{AQ}{Z}\right)$$

$= \left(\frac{P-AQ}{Z}\right)$; but this is $\frac{R}{Z}$, which is therefore rational, or Z also rationally divides R. By similar reasoning the other cases follow.

7. The highest common divisor of P and Q is therefore the highest common divisor of Q and R.

* Not strictly a *quotient*, unless the remainder be nothing. It is what comes in the quotient-part of the process in trying this point.

8. If one factor of a product be not divisible by x, then those powers of x, and those only, which rationally divide the other factor, will rationally divide the product. For instance, in $(x^4+a)(x^3+6x^2)$, since the lowest term is $6ax^2$, and since no power of x higher than that in the lowest term will rationally divide an expression, it follows that x^2 is the highest power of x which rationally divides the product. But x^2 is the highest which rationally divides x^3+6x^2.

9. If an expression be rationally divisible by a power of x, the quotient is divisible by every divisor of the first, which is itself indivisible by any power of x. For example, x^3-x divided by x gives x^2-1, $x-1$ is a divisor of the first, and, therefore (were it not known otherwise), is a divisor of the second.

To prove this theorem, let x^3P (for example) be an expression, which is divisible, say by $x+1$; that is, let

$$\frac{x^3(P)}{x+1} = (A) \qquad\qquad x^3(P) = (A)(x+1)$$

consequently (8.), (A) is divisible by x^3;

or $$\frac{A}{x^3} = (B) \quad \text{that is} \quad (A) = x^3(B)$$

Therefore, $$x^3(P) = x^3(B)(x+1)$$
$$(P) = (B)(x+1) \quad \text{or} \quad \frac{P}{x+1} = (B)$$

that is, P (as well as x^3P) is divisible by $x+1$, and the same may be shewn of any other divisor of x^3P.

The method of finding the highest common divisor of two rational polynomials is now exactly similar to that of finding the greatest common measure of two whole numbers in arithmetic. For example, required the highest common divisor of x^6-x and $3x^8-3x^4$. First separate the monomial factors; that is, put the expressions in the form

$$x(x^5-1) \quad \text{and} \quad 3x^4(x^4-1)$$

Neglect the monomial factors for the present, and proceed to find the highest common divisor of

$$x^5-1 = P \quad \text{and} \quad x^4-1 = Q$$

$$x^4-1)x^5-1(x$$

$$\text{Rem.} \qquad x-1)x^4-1(x^3+x^2+x+1$$

$$\text{Rem. } 0$$

By (7.) the highest divisor of x^5-1 and x^4-1 is also that of x^4-1 and $x-1$; and, since $x-1$ divides the former, and also is the highest divisor of itself, it is the highest divisor common to the two, and, therefore, the highest divisor of x^5-1 and x^4-1. The original expressions have also the divisor x; consequently, $x(x-1)$ is their highest common divisor.

Any power of x may be thrown out of a remainder by division, from theorem 9. For instance, in finding the highest common divisor of $1-x^3$ and $1-x^5$; the first remainder is x^3-x^5 or $x^3(1-x^2)$; but $1-x^2$ has all the divisors of x^3-x^5, except those which are powers of x. But $1-x^5$ and $1-x^3$ are neither divisible by a power of x; consequently, $1-x^2$ contains all their common divisors, as well as x^3-x^5, and the former may be used for the latter in any division.

The divisor may be taken any number of times which may be convenient, before using it as a new dividend. For instance, in finding the greatest common measure of x^2-2x+1 and x^3-1, the first remainder is $2x^2-x-1$; before dividing by this,* it will be convenient to multiply x^2-2x+1 by 2, and no new *common* divisor will thus be introduced. The whole process in the latter case may be as follows, which, though not the shortest possible, will illustrate the methods to be applied to more complicated cases.

$$x^2-2x+1)\,x^3-1\,(x$$
$$x^3-2x^2+x$$
$$2x^2-x-1)\;2x^2-4x+2\,(1$$
$$2x^2-x-1$$
$$-3x+3$$

Divide by -3

$$x-1)\,2x^2-x-1\,(2x+1$$
$$2x^3-2x$$
$$x-1$$
$$x-1$$
$$0$$

Therefore $x-1$ is the highest common divisor.

* The division might be carried one step further before using the remainder; but either method answers equally well.

CHAPTER VIII.

ON SERIES AND INDETERMINATE COEFFICIENTS.

WE have already seen (page 160) that the sum of the terms (x being less than 1)

$$1 + x + x^2 + x^3 + x^4 + \text{ \&c.}$$

how far soever it may be carried, never can exceed or come up to $1 \div (1 - x)$. This expression is called an infinite series, and the fraction $1 \div (1 - x)$ which (page 150) might be called the *limit* of the sum, is called the *sum*.

Definition. By the *sum of an infinite series* is meant the limit towards which we approximate by continually adding more and more of its terms.

A *convergent* series is one in which such a limit exists, that is, in which we cannot attain a number as great as we please by summing its terms: a *divergent* series is one in which there is no limit to the quantity which may be attained by summing its terms. The following series are divergent, and all but the last, evidently so.

$$1 + 1 + 1 + 1 + \text{ \&c.} \qquad 1 + 2 + 3 + 4 + \text{ \&c.}$$

$$1 + 2 + 4 + 8 + \text{ \&c.} \qquad 1 + \frac{1}{2} + \frac{1}{3} + \frac{1}{4} + \text{ \&c.*}$$

In an infinite series, we must know the connexion which exists between each term and the next, otherwise we cannot reason upon it. For, as we cannot write down all the terms, it is only from knowing the connexion between successive terms that we can be said to know of what series we are speaking. So that an infinite series with no *law of connexion* existing between its terms, has no existence for the purposes of reasoning.

The student might perhaps imagine that the law is immediately perceptible when the first four or five terms are given, and an obvious connexion exists between them. For instance, he would suppose that the following series

$$1 + 1 + 1 + 1 + 1 + 1 + 1 + 1 + 1 + \text{ \&c.}$$

* See *Arithmetic,* article 197.

if it be to be continued according to a law actually existing among the given terms, must be a succession of units added together. But this is not the case; the preceding series might be continued in an infinite number of different ways, each different method following a law which actually exists among the given terms as they stand. For instance, the preceding series might be thus continued:

9th term.	10th.	11th.	12th.	13th.	14th.	15th.	16th.	
1	1	2	3	4	5	7	10	&c.

the law of which is, that the $(n+1)$th exceeds the nth by the *tens figure* of the sum of the first n terms, which requires that the terms should remain equal until their sum has a figure in the second column. The following series of terms have laws which we leave to the reader to detect.

7	16	22	26	32	36	42	*ad infin.*			
5	10	9	10	9	10	9	*ad infin.*			
5	10	11	15	21	30	39	43	52	61	70
79	85	94	103	109	109	109	*ad infin.*			

If we attempt to deduce general propositions from a few particular cases, as, for instance, the law of a series from that of a few of its terms, we are liable to error. All that can be derived from observing a few cases is a strong presumption, high probability, or great likelihood, that the law observed is always true. But that which is beforehand very likely, does not always turn out, on examination, to be true: as in the following instance. Take the series of numbers 1, 2, 3, 4, 5, &c. multiply each by the next higher, and add 41 to the product, as follows:

$1 \times 2 + 41 = 43$ $\quad$ $5 \times 6 + 41 = 71$

$2 \times 3 + 41 = 47$ $\quad$ $6 \times 7 + 41 = 83$

$3 \times 4 + 41 = 53$ $\quad$ $7 \times 8 + 41 = 97$

$4 \times 5 + 41 = 61$ $\quad$ $8 \times 9 + 41 = 113$, &c.

On examining the series of results

43, 47, 53, 61, 71, 83, 97, 113, &c.

we see that all of them seem to be *prime** numbers, and hence we

* A prime number is one which does not admit of any divisor except 1, and itself. The series of prime numbers is

1, 2, 3, 5, 7, 11, 13, 17, 19, &c.

have a very strong reason for presuming that this will continue to be the case, that is, we suspect the following to be true: if x be any whole number, $x(x+1)+41$ is a prime number. And on continuing the series, we actually find prime numbers, and nothing but prime numbers, up to $39 \times 40 + 41$ or 1601. But, nevertheless, the next term, or $40 \times 41 + 41$, is evidently not a prime number; for it is $(40+1)41$, or 41×41.

To avoid the continual necessity of expressing the law of a series, we always mean, in future, that, where a simple law appears among the few terms which are written down, that law is to be the law of the series, unless some other law be mentioned. Thus $1+x+x^2+$ &c. implies that the succeeding terms are $x^3+x^4+x^5+$ &c.

DEFINITION. The *general term* of a series is the algebraical expression for the nth term, as will be better understood from the following cases.

First few terms.	nth, or general term.
$1+1+1+1+$ &c.	1
$1+2+3+4+$ &c.	n
$2+3+4+5+$ &c.	$n+1$
$0+1+2+3+$ &c.	$n-1$
$1+4+9+16+$ &c.	n^2
$4+9+16+25+$ &c.	$(n+1)^2$
$x+x^2+x^3+x^4+$ &c.	x^n
$1+x+x^2+x^3+$ &c.	x^{n-1}
$x^m+x^{m+1}+x^{m+2}+x^{m+3}+$ &c.	x^{m+n-1}
$1+\frac{x}{2}+\frac{x^2}{3}+\frac{x^3}{4}+$ &c.	$\frac{x^{n-1}}{n}$
$1+x+\frac{x^2}{2}+\frac{x^3}{2.3}+$ &c.	$\frac{x^{n-1}}{1.2.3\ldots\ldots(n-1)}$

In the last series, the first term is not included in the general term, as given. For if $n=1$, the general term becomes $\frac{x^{1-1}}{0}$ or $\frac{1}{0}$, which is not true. Properly speaking, the general term is n factors of the following product:

$$1 \times \frac{x}{1} \times \frac{x}{2} \times \frac{x}{3} \times \frac{x}{4} \ldots\ldots\ldots\ldots$$

THEOREM. The series $a+b+c+e+f+$ &c. is the same as the following:

$$a\left\{1+\frac{b}{a}+\frac{c\,b}{b\,a}+\frac{e\,c\,b}{c\,b\,a}+\frac{f\,e\,c\,b}{e\,c\,b\,a}+\text{\&c.}\right\}$$

The student will have no difficulty in proving this. Let the ratio* of each term to the preceding term be denoted by the capital letter of the numerator.

$$\frac{b}{a}=\mathrm{B}\quad \frac{c}{b}=\mathrm{C}\quad \frac{e}{c}=\mathrm{E}\quad \frac{f}{e}=\mathrm{F}\ \text{\&c.}$$

Then $a+b+c+e+f+$ &c. is

$$a\left\{1+\mathrm{B}+\mathrm{CB}+\mathrm{ECB}+\mathrm{FECB}+\text{\&c.}\right\}\cdots\cdots(1)$$

If every one of the ratios, B, C, E, F, &c., be less than some given quantity, say P, then

$$a(1+\mathrm{B}) \quad \text{is less than} \quad a(1+\mathrm{P})$$
$$a(1+\mathrm{B}+\mathrm{CB}) \cdots\cdots\cdots \ a(1+\mathrm{P}+\mathrm{PP})$$
$$\text{\&c.} \qquad\qquad \text{\&c.}$$

or the sum of any number of terms of (1) is less than that of the same number of terms of

$$a(1+\mathrm{P}+\mathrm{P}^2+\mathrm{P}^3+\text{\&c.}) \cdots\cdots\cdots (2)$$

If, then, P itself be less than unity, (1) must be a convergent series; for no number of terms of $1+\mathrm{P}+\mathrm{P}^2+$ &c. can then exceed $1 \div (1-\mathrm{P})$, consequently, no number of terms of (2) can exceed $a \div (1-\mathrm{P})$; still less can any number of terms of (1) exceed the same, because the terms of (1) are severally less than those of (2).

Consequently, *a series is always convergent when the ratio of* any *term to the preceding term is less than some quantity, which is itself less than unity.* It is sufficient that this should happen after some certain number of terms: for, say that the first hundred terms are increasing terms, yet if no summation of terms after the hundredth will give a result exceeding, say 50, and if the sum of the first hundred terms be, say 1000, then no summation whatever will give a result exceeding 1050, or the series is on the whole convergent, or, properly speaking, begins to converge after the hundredth term.

* $\frac{a}{b}$ is the algebraical synonyme for what is called in Euclid "the ratio of a to b," and the geometrical term, which is a highly convenient one, is frequently adopted.

Example. $$1+1+\frac{1}{2}+\frac{1}{2.3}+\frac{1}{2.3.4}+\&c.$$
is convergent. For here we have
$$\frac{b}{a}=1 \quad \frac{c}{b}=\frac{1}{2} \quad \frac{e}{c}=\frac{1}{3} \quad \frac{f}{e}=\frac{1}{4}\ \&c.$$
so that each ratio after the second is less than $\frac{1}{2}$, which is less than unity.

[As the limit of this series is an important number in algebra, we shall proceed to find it as far as 10 decimal places, using eleven places to insure the accuracy of the 10th. Let the terms be called a_1, a_2, &c., then we have
$$a_1=1 \quad a_2=1 \quad a_3=\frac{1}{2}a_2 \quad a_4=\frac{1}{3}a_3 \quad a_5=\frac{1}{4}a_4\ \&c.$$

$a_1=1$	....	1·00000000000
$a_2=1$	....	1·00000000000
$a_3=\frac{1}{2}a_2$	....	0·50000000000
$a_4=\frac{1}{3}a_3$	....	·16666666667
$a_5=\frac{1}{4}a_4$	....	·04166666667
$a_6=\frac{1}{5}a_5$	....	·00833333333
$a_7=\frac{1}{6}a_6$	....	·00138888889
$a_8=\frac{1}{7}a_7$	....	·00019841270
$a_9=\frac{1}{8}a_8$	....	·00002480159
$a_{10}=\frac{1}{9}a_9$	....	·00000275573
$a_{11}=\frac{1}{10}a_{10}$	....	·00000027557
$a_{12}=\frac{1}{11}a_{11}$	....	·00000002505
$a_{13}=\frac{1}{12}a_{12}$	....	·00000000209
$a_{14}=\frac{1}{13}a_{13}$	....	·00000000016
$a_{15}=\frac{1}{14}a_{14}$	....	·00000000001
		2·71828182846

This is correct to the last place; in fact, the sum of the series lies between

$$2{\cdot}71828182845$$

and $$2{\cdot}71828182846$$

but nearer to the latter. The letter ε (and sometimes e) is used to denote the limit of this sum; or we say that

$$\varepsilon = 1 + 1 + \frac{1}{2} + \frac{1}{2.3} + \&c. \ (= 2{\cdot}71828182846 \text{ very nearly.})$$

The above series is said to *converge rapidly*.]

THEOREM. The series $a + b + c +$ &c. is always divergent, whenever its terms are so related that $\frac{b}{a}, \frac{c}{b}$, &c., are all greater than unity, or continue so from and after a given term.

As the demonstration of this theorem is very like that of the last, we leave it to the student.

THEOREM. The series $a - b + c - e +$ &c. is convergent whenever the terms decrease without limit; that is, when a is greater than b, b greater than c, &c., and when some term or other of the series must be less than any fraction we may name.

Let a, b, c, e, &c. be a series of decreasing terms, as in the theorem, then

$$(a - b) + (b - c) + (c - e) + \&c.$$

must be a converging series, for the sum of the first two terms is $a - c$, of the first three, $a - e$, and so on. Now, since a, b, c, e .... decrease without limit, $a - c$, $a - e$, &c. is a series of increasing terms which has the limit a. Consequently, the series made by taking only alternate terms of the preceding, must have a limit less than a. But that series is

$$a - b + c - e + \&c.$$

whence the theorem is proved.

Hence we know that

$$1 - \frac{1}{2} + \frac{1}{3} - \frac{1}{4} + \&c.$$

is convergent, with a limit less than 1.

THEOREM. If any given quantity P be greater than any one of the series of ratios

$$\frac{b}{a} \ \frac{c}{b} \ \frac{e}{c} \ \frac{f}{e} \ \frac{g}{f} \quad \&c.$$

then the series

$$a+bx+cx^2+ex^3+fx^4+gx^5+ \text{\&c.} \ldots. \text{(A)}$$

is convergent whenever x is less than $\frac{1}{P}$.

For, since the preceding series is

$$a\left\{1+\frac{b}{a}x+\frac{c}{b}\cdot\frac{b}{a}\cdot x^2+\frac{e}{c}\cdot\frac{c}{b}\cdot\frac{b}{a}x^3+\text{\&c.}\right\}$$

and since P is greater than $\frac{b}{a}$, $\frac{c}{b}$, &c., the preceding series will be increased by writing P instead of $\frac{b}{a}$, $\frac{c}{b}$, &c. But it then becomes

$$a\left\{1+Px+P^2x^2+P^3x^3+\text{\&c.}\right\}$$

or

$$a\left\{1+(Px)+(Px)^2+(Px)^3+\text{\&c.}\right\}$$

which (page 159) is convergent if Px be less than 1, or x less than $\frac{1}{P}$. Still more is the original series convergent under the same circumstances, because its terms are severally less than those of the last series.

If P be greater than any one of the ratios after some given ratio, the series converges from and after the term which gives that ratio, whenever Px is less than 1. Suppose, for instance, that the thousandth and following terms of the series (A) are

$$Ax^{999}+Bx^{1000}+Cx^{1001}+\text{\&c.}$$

or

$$Ax^{999}\left\{1+\frac{B}{A}x+\frac{C}{B}\frac{B}{A}x^2+\text{\&c.}\right\}$$

Then, by the preceding reasoning, if P be greater than any of the ratios $\frac{B}{A}$, $\frac{C}{B}$, &c. the preceding series converges if x be less than $\frac{1}{P}$.

For instance, take

$$1+2x+3x^2+4x^3+\text{\&c.}$$

ratios $\quad \frac{2}{1} \quad \frac{3}{2} \quad \frac{4}{3} \quad \frac{5}{4} \quad$ &c.

2 is greater than any of these ratios after the first; consequently, this series converges from the second term if x be less than $\frac{1}{2}$. The hundredth and following terms are

$$100x^{99}+101x^{100}+102x^{101}+\text{\&c.}$$

ratios $\quad \frac{101}{100} \quad \frac{102}{101} \quad \frac{103}{102} \quad$ &c.

of which $\frac{101}{100}$ is greater than any except the first. Consequently, this series converges from the hundredth term, if x be less than $1 \div \frac{101}{100}$ or $\frac{100}{101}$. Similarly it may be shewn that this series is convergent whenever x is less than 1, though the term at which convergency begins may be made as distant as we please, by making x sufficiently near to 1.

As a second example, take

$$1 + x + \frac{x^2}{2} + \frac{x^3}{2.3} + \frac{x^4}{2.3.4} + \&c.$$

ratios $\quad 1 \quad \frac{1}{2} \quad \frac{1}{3} \quad \frac{1}{4} \quad$ &c.

Since these ratios continually diminish, and without limit, a point of the series will come, after which they will all be less than any given fraction m, however small it may be. But if m may be made as small as we please, $\frac{1}{m}$ may be made as great as we please. Therefore this series is convergent for every value of x, however great; though the greater x is taken the more distant will be the term at which convergency begins.

As a third example, take

$$1 + 2x + 2.3\,x^2 + 2.3.4\,x^3 + \&c.$$

ratios $\quad 2, \quad 3, \quad 4, \quad 5, \quad$ &c.

and as these ratios increase without limit, there is no quantity which is greater than them all. Consequently, no value can be assigned to x for which this series must necessarily be convergent. The following theorem may easily be proved in the same manner as the last.

THEOREM. If P be less than any one of the ratios $\frac{b}{a}$ $\frac{c}{b}$, &c.; then the series

$$a + bx + cx^2 + \&c.$$

must be divergent for every value of x greater than $\frac{1}{P}$.

In this way the series in the last example may be shewn to diverge from the second term for every value of x greater than $\frac{1}{3}$, from the third for every value greater than $\frac{1}{4}$, and so on; so that there is no fraction so small that the series shall not diverge from and after some term by giving x that value.

[Such necessarily divergent series never will be found in practice: they are introduced here as a warning against applying to *all* series general conclusions drawn from series which may be made convergent.]

In future, unless the contrary be specially mentioned, in speaking of the series $a + bx + cx^2 +$ &c., *we only mean to speak of series which may be made convergent. We suppose all the terms positive.*

THEOREM. Every series of the form $a + bx + cx^2 +$ &c. has this property, that x may be taken so small, that any one term shall contain the aggregate of all the following terms as often as we please.

For instance, by taking x sufficiently small, we may make cx^2 more than ten thousand times $ex^3 + fx^4 +$ &c. Let x_1 be the greatest value of x which makes $e + fx +$ &c. convergent, and let the sum in that case be S. Then, for every value of x less than x_1, $e + fx +$ &c. is less than S. Now, cx^2 contains $ex^3 + fx^4 +$ &c.

$$\frac{cx^2}{ex^3 + fx^4 + \text{\&c.}} \quad \text{or} \quad \frac{c}{ex + fx^2 + \text{\&c.}} \quad \text{or} \quad \frac{c}{x(e + fx + \text{\&c.})}$$

times or parts of times. Take x less than x_1, so that S is greater than $e + fx +$ &c., or

$$\frac{c}{xS} \quad \text{less than} \quad \frac{c}{x(e + fx + \text{\&c.})} \quad \text{or} \quad \frac{cx^2}{ex^3 + fx^4 + \text{\&c.}}$$

Now, c and S being fixed quantities, x may be taken so small that $c \div xS$ shall be as great as we please; and still more $cx^2 \div (ex^3 + fx^4 +$ &c.), which is greater than $c \div xS$. Whence the theorem is proved.

EXAMPLE. How small must x be taken, so that we may be sure the fourth term of

$$1 + 2x + 3x^2 + 4x^3 + 5x^4 + \text{\&c.}$$

contains the sum of all that follow 1000 times at least.

The whole of the series after the fourth term may be written thus:

$$5x^4 \left\{1 + \frac{6}{5}x + \frac{7}{6}\cdot\frac{6}{5}x^2 + \text{\&c.}\right\} \ldots\ldots\ldots\ldots\ldots (\text{A})$$

and $\frac{6}{5}$ is greater than any of the succeeding ratios; consequently, we increase the preceding by altering it to

$$5x^4\left\{1+\frac{6}{5}x+\frac{6}{5}\cdot\frac{6}{5}x^2+\text{\&c.}\right\} \text{ or } \frac{5x^4}{1-\frac{6}{5}x} \ldots\ldots (B)$$

(see page 182.) We have then to take x, so that

$$4x^3 \text{ is greater than } 1000\frac{5x^4}{1-\frac{6}{5}x}$$

$(\times)\left(\frac{1-\frac{6}{5}x}{4x^3}\right)$, $1-\frac{6}{5}x$ must be greater than $250\times 5x$ or $1250x$, which is certainly true if $1-2x$ be greater than $1250x$, or 1 greater than $1252x$, or $\frac{1}{1252}$ greater than x. In this case $4x^3$ is greater than 1000 times (B); still more then is it greater than 1000 times (A).

Theorem. If the two series

$$a_0+a_1x+a_2x^2+\text{\&c.} \text{ and } b_0+b_1x+b_2x^2+\text{\&c.}$$

be *always* equal for every finite value of x, then it must follow that $a_0=b_0$, $a_1=b_1$, $a_2=b_2$, &c. or the series are identically the same.

Let these series be called a_0+A and b_0+B, in which, by what has just been proved, we can make A and B less than the mth parts of a_0 and b_0. If possible, let a_0 and b_0 be different numbers, and let $a_0=b_0+t$. Then, since the series are always equal, we have $a_0+A=b_0+B$ or $b_0+t+A=b_0+B$; that is, $t=B-A$. But because a_0 and b_0 are fixed quantities, their difference is the same; and we have t, a fixed quantity, equal to the difference of two quantities, each of which may be made as small as we please, which is absurd. Hence $a_0=b_0+t$ cannot be; and by the same reasoning, $a_0=b_0-t$ cannot be; therefore $a_0=b_0$. Take away these equal terms from the two equal series, and divide the equal remainders by x, which gives

$$a_1+a_2x+a_3x^2+\text{\&c. always equal to } b_1+b_2x+b_3x^2+\text{\&c.}$$

from which the same species of proof gives $a_1=b_1$; repeat the process of subtracting equal terms and dividing by x, and the repetition of the proof gives $a_2=b_2$; and so on. Hence, if one or more terms be wanting in either series, the same must be wanting in the other; for instance, if $a-x$ be always equal to $a_0+a_1x+a_2x^2+$ &c., we must have $a=a_0$, $-1=a_1$, $0=a_2$, $0=a_3$, &c.

[The preceding process amounts simply to shewing that we may make $x = 0$ in a series, and take the ordinary algebraical consequences, in the same manner as if the expression were finite in its number of terms. For instance, if $a_0 + a_1 x +$ &c. be always $= b_0 + b_1 x +$ &c., *we have proved* that the consequence of making $x = 0$, namely, $a_0 = b_0$, is true. But, as we have sufficiently seen, it is not safe to say that when $x = 0$, P = Q, except in cases where we may say that by making x sufficiently small (or near to nothing), P may be brought as near to Q as we please. The difficulty which we have avoided is as follows: In the series $1 + 2x + 2.3x^2 +$ &c. we have seen that, take x as small as we may, the sum of the terms can be made as great as we please. Are we, then, entitled to say, that when $x = 0$, the preceding becomes $1 + 0 + 0 +$ &c. or 1? If the number of terms were finite, there could be no doubt of the propriety of answering in the affirmative; but when the number of terms is infinite, nothing that has preceded will enable us to give an answer. The student will remember that we have confined the demonstration entirely to series which admit of being made convergent.

It is usual to prove the preceding* by saying, that when the two series are always equal, they are equal when $x = 0$, and consequently $a_0 = b_0$, and so on. This is avoided in the present case; and we may say that we have proved the following theorem. If two series (which can be made convergent) are always equal when x is finite, then they are also equal when $x = 0$.]

* On this point the student, when he is more advanced, may consult Professor Woodhouse, *Analytical Calculations*, &c. Preface, p. viii. note. It is there objected that to make $x = 0$ and thence to deduce $a_0 = b_0$, is the same as arbitrarily making $a_0 = b_0$. This I conceive to be not the true point of difficulty. All mathematical consequences are necessarily contained in the hypotheses from which they spring: so that to invent any hypothesis is necessarily to invent all its consequences, some of which may be so near as to appear nothing more than the hypothesis itself, others so little perceptible as not to seem necessary attendants of the hypothesis. The real objection to the proof on which this note is written, I conceive, is this, that having frequently found the passage from x as a symbol of magnitude, to x as the symbol not of magnitude but of the absence of all magnitude, to be attended with consequences which require a special examination, it is not allowable to enter upon any new ground, without either establishing the accordance of the consequences of $x = 0$, with those of $x =$ some magnitude, or distinguishing and explaining the discrepancy, if any.

The following are a few instances of the method by which we can obtain the limits of the sums of many series. First, let us take

$$P = 1+x+x^2+x^3+x^4+ \text{\&c.}$$

in which we wish to determine a finite algebraical expression for P. It is plain that

$$1+x+x^2+ \text{\&c. } ad\ inf. = 1+x\left\{1+x+x^2+ \text{\&c. } ad\ inf.\right\}$$

$$\text{that is,}\quad P = 1+xP \quad \text{or} \quad P = \frac{1}{1-x}$$

a result previously ascertained. Now, let us take

$$P = 1+2x+3x^2+4x^3+ \text{\&c.}$$

$$\frac{P-1}{x} = 2+3x+4x^2+5x^3+ \text{\&c.}$$

$$\frac{P-1}{x} - P = 1+x+x^2+x^3+ \text{\&c.} = \frac{1}{1-x}$$

whence $$P\left\{\frac{1}{x}-1\right\} = \frac{1}{1-x}+\frac{1}{x} = \frac{1}{x}\cdot\frac{1}{1-x}$$

or $$P = \frac{1}{(1-x)^2}$$

Next, let $$P = 1+3x+5x^2+7x^3+ \text{\&c.}$$

$$\frac{P-1}{x} = 3+5x+7x^2+9x^3+ \text{\&c.}$$

$$\frac{P-1}{x} - P = 2+2x+2x^2+2x^3+ \text{\&c.} = \frac{2}{1-x}$$

whence $$P = \frac{1+x}{(1-x)^2}$$

Next, let $$P = 1+4x+9x^2+16x^3+ \text{\&c.}$$

$$\frac{P-1}{x} = 4+9x+16x^2+25x^3+ \text{\&c.}$$

$$\frac{P-1}{x} - P = 3+5x+7x^2+9x^3+ \text{\&c.}$$

$$\therefore\quad \left(\frac{P-1}{x} - P\right)x+1 = 1+3x+5x^2+ \text{\&c.} = \frac{1+x}{(1-x)^2}$$

whence $$P = \frac{1+x}{(1-x)^3}$$

The same method might be applied to finding a more simple algebraical expression for the sum of any finite number of terms of the preceding series. For example, let

$$P = 1+2x+3x^2+ \;\cdots\cdots\; +(n-1)x^{n-2}+nx^{n-1}$$

$$\frac{P-1}{x} = 2+3x+4x^2+ \;\cdots\cdots\; +nx^{n-2}$$

$$\frac{P-1}{x}-P = 1+x+x^2+ \;\cdots\cdots\; +x^{n-2}-nx^{n-1}$$

$$\text{(page 103)} = \frac{1-x^{n-1}}{1-x}-nx^{n-1} = \frac{1-(n+1)x^{n-1}+nx^n}{1-x}$$

$$P = \frac{nx^{n+1}-(n+1)x^n+1}{(1-x)^2}$$

The student may endeavour to prove the following:

$$1+3x+5x^2+ \ldots +(2n-1)x^{n-1} = \frac{\overline{2n-1}\,x^{n+1}-\overline{2n+1}\,x^n+x+1}{(1-x)^2}$$

The inquiry which we have most frequently to make is the inverse of the preceding; not, having given the series to find its sum, but, having given an expression, to find the series of which it is the sum, or to *develope it in powers of some one of the letters contained in it.* Let us suppose, for example, that we want a series of powers of x with coefficients, &c. which shall be, in all cases in which it is convergent, equal to $(1+x) \div (1-x)^2$. Suppose the series to be $a_0+a_1x+a_2x^2+$ &c., so that we have

$$\frac{1+x}{(1-x)^2} = a_0+a_1x+a_2x^2+a_3x^3+\text{ \&c.}$$

Multiply both sides by $(1-x)^2$ or $1-2x+x^2$

$$1+x = \left\{\begin{array}{r} a_0+a_1\,x+a_2\,x^2+a_3\,x^3+\text{ \&c.} \\ -2a_0x-2a_1x^2-2a_2x^3-\text{ \&c.} \\ +a_0\,x^2+a_1\,x^3+\text{ \&c.} \end{array}\right\}$$

$$= a_0+(a_1-2a_0)x+(a_2-2a_1+a_0)x^2+\text{ \&c.}$$

The two sides of this equation being equal for every value of x, the theorem in page 188 gives

$$\begin{array}{llll} a_0=1 & a_1-2a_0=1 & \text{or} & a_1=3 \\ & a_2-2a_1+a_0=0 & \text{or} & a_2=5 \\ & a_3-2a_2+a_1=0 & \text{or} & a_3=7 \\ & \text{\&c.} & & \text{\&c.} \end{array}$$

So that the series is $1+3x+5x^2+$ &c. as already determined.

The first term of the preceding might have been found immediately: for, since we have proved that the results of $x=0$ may be employed, and since $(1+x)\div(1-x)^2$ becomes 1, and the series is reduced to a_0, when $x=0$, we have $a_0=1$.

We shall now ask what is $(1-x^4)\div(1-x)$ expanded in a series of powers of x. The first term is 1, found as in the last entence; let us suppose

$$\frac{1-x^4}{1-x} = 1+a_1x+a_2x^2+a_3x^3+a_4x^4+ \&c.$$

$$1-x^4 = 1+a_1x+a_2x^2+a_3x^3+a^4x^4+a_5x^5+ \&c.$$
$$- \quad x-a_1x^2-a_2x^3-a_3x^4-a_4x^4+ \&c.$$

As there is no first power of x on the first side, we must have $a_1-1=0$, or $a_1=1$. Similarly, $a_2-a_1=0$, or $a_2=a_1=1$; $a_3-a_2=0$, $a_3=1$. But the coefficient of x^4 on the first side being -1, we must have $a_4-a_3=-1$, or $a_4-1=-1$, that is, $a_4=0$; again, $a_5-a_4=0$, or $a_5=0$, $a_6-a_5=0$, or $a_6=0$, and so on. Hence the series is

$$1+x+x^2+x^3+0\times x^4+0\times x^5+ \&c.$$

that is $\quad 1+x+x^2+x^3$

as might be found by simple division, or from page 103. Thus, we see that when we assume an infinite series to represent a quantity which is in fact a finite expression, the method of determining the coefficients of the series will shew the coefficient 0 for every term of the series which does not exist in the finite expression.

Again, to develope $1\div(1+x^2)$ assume

$$\frac{1}{1+x^2} = a_0+a_1x+a_2x^2+a_3x^3+ \&c.$$

the preceding process will give

$$a_0=1 \quad a_2+a_0=0 \text{ or } a_2=-1 \quad a_4+a_2=0 \text{ or } a_4=1$$
$$a_1=0 \quad a_3+a_1=0 \text{ or } a_3=0 \quad a_5+a_3=0 \text{ or } a_5=0$$

so that the series is

$$1+0\times x-x^2+0\times x^3+x^4+0\times x^5 \&c.$$

or $\quad 1-x^2+x^4-x^6+ \&c.$

If it happen that, by the preceding process, an equation is produced of which the two sides cannot be made identical by any suppositions as to the value of the coefficients, it is a sign that the expression cannot be developed in a series of the form proposed. If we try to develope $1\div(1+x)x$ by assuming

$$\frac{1}{x(1+x)} = a_0 + a_1 x + a_2 x^2 + a_3 x^3 + \text{\&c.}$$

we shall find

$$1 = a_0 x + (a_1 + a_0) x^2 + (a_2 + a_1) x^3 + \text{\&c.}$$

the two sides of which cannot on any supposition be made to agree; for there is no term independent of x on the second side which may be made $= 1$. In fact, we should find

$$\frac{1}{1+x} = 1 - x + x^2 - x^3 + \text{\&c.}$$

$$(\div)\, x \qquad \frac{1}{x(1+x)} = \frac{1}{x} - 1 + x - x^2 + \text{\&c.}$$

so that the fraction proposed cannot be developed entirely in whole and *positive* powers of x, without the introduction of a negative power (in this case x^{-1}).

The following are given as exercises:

1. If $P = a_0 + a_1 x + a_2 x^2 + \text{\&c.}$ then

$$\frac{P}{1-x} = a_0 + (a_0 + a_1) x + (a_0 + a_1 + a_2) x^2 + \text{\&c.}$$

$$\frac{P}{1+x} = a_0 + (a_1 - a_0) x + (a_2 - a_1 + a_0) x^2 + \text{\&c.}$$

2. $$\frac{1}{1+x+x^2} = 1 - x + x^3 - x^4 + x^6 - x^7 + \text{\&c.}$$

3. $$\frac{1+cx}{c+x} = \frac{1}{c} - \left(\frac{1}{c^2} - 1\right) x + \left(\frac{1}{c^3} - \frac{1}{c}\right) x^2 - \text{\&c.}$$

4. $$\frac{m+nx}{p+qx} = \frac{m}{p} - \left(\frac{mq}{p^2} - \frac{n}{p}\right) x + \left(\frac{mq^2}{p^3} - \frac{nq}{p^2}\right) x^2 - \text{\&c.}$$

CHAPTER IX.

ON THE MEANING OF *EQUALITY* IN ALGEBRA, AS DISTINGUISHED FROM ITS MEANING IN ARITHMETIC.

In page 62, among the extensions of terms, we notified that the word equal was to be considered as applicable to any two expressions of which one could be substituted for the other without error. Hitherto we have only applied this extension to the case of definite algebraical quantities, either positive or negative: the numerical value of the quantity determining its magnitude, the sign determining only which of two opposite relations is intended to be expressed. We now proceed to consider the word equal, or its sign $=$, not in a sense wider than any which the definition will bear, but wider than any in which we have yet had occasion to use it.

Two expressions are said to be equal when one can be substituted for the other without error. The whole force of this definition lies in the answer to the question, What is error? The answer is, any thing which leads to contradictory results, or which may in any legitimate way be made to lead to contradictory results.

Results are contradictory when, both being intelligible, or capable of interpretation, the two do not agree; but they are not necessarily contradictory merely because one or both are unintelligible, for where the meaning of any part of an assertion is unknown, we cannot say whether it be true or false. For instance, in page 23, the expression $x = \frac{c}{0}$ was no contradiction of any thing which had preceded, for $\frac{c}{0}$ had no meaning. Having ascertained this, our object was to give it a meaning which should not contradict any thing deducible from preceding principles.

We have shewn that if x be less than unity, the summation of $1 + x + x^2 +$ &c. will continually give results nearer to $\frac{1}{1-x}$, which however can never be absolutely reached by that process. Hence we used the sign $=$ in the equation

$$\frac{1}{1-x} = 1+x+x^2+x^3+ \text{\&c. } ad\ infinitum.$$

Arithmetically speaking, we can make the preceding as nearly true as we please, by taking a sufficient number of terms from the second side. Algebraically speaking, the above may be considered as absolutely true; but it must be remembered that this is only on a supposition which is arithmetically impossible, namely, that all the terms on the second side, *expressed* and *implied*, shall be considered as included. For instance, we know that the multiplication of the first side by $1-x^2$ gives $1+x$; the multiplication of the second side by the same gives

$$\begin{cases} 1+x+x^2+x^3+x^4+x^5+ \text{\&c. } ad\ inf. \\ \qquad\quad -x^2-x^3-x^4-x^5- \text{\&c. } ad\ inf. \end{cases}$$

or $$1+x+0+0+0+0+ \text{\&c. } ad\ inf.$$

in which we can prove, if necessary, that every succeeding term shall be 0; not by actually looking at every term, which is impossible, but by a deduction from what we know to be the law of the series. Similarly, we shew that $x^m \times x^n = x^{m+n}$; not by looking at every case, which is impossible, but by what we know of the meaning of x^m.

If we proceed arithmetically, taking any number of terms, however great, say as far as x^n, we then have

$$\begin{aligned} &(1+x+x^2+ \ \ldots\ +x^{n-1}+x^n)\times(1-x^2) \\ = &\begin{cases} 1+x+x^2+x^3+ \ \ldots\ +x^{n-1}+x^n \\ \qquad\quad -x^2-x^3- \ \ldots\ -x^{n-1}-x^n-x^{n+1}-x^{n+2} \end{cases} \\ = &\ 1+x-x^{n+1}-x^{n+2} \end{aligned}$$

in which, since x is less than 1, n can be taken so great that x^{n+1} and x^{n+2} shall be as small as we please. Here is algebraical equality which can be made arithmetical equality, *quam proximè*.

Let us now suppose that x is greater than 1; say $x=2$; the series

$$1+x+x^2+x^3+ \text{\&c. } ad\ infinitum$$

is then, arithmetically speaking, infinite, since there is no limit to the magnitude we may obtain by summing $1+2+4+8+$ &c. Should we then assert algebraical equality between $\frac{a}{0}$ and the preceding series? We have no right to do so from pages 21, &c. for though it was there shewn how to make it clear that $\frac{a}{0}$ is above all arithmetical numeration, the converse does not therefore follow, that whatever is above arithmetical numeration is properly represented by $\frac{a}{0}$. What, then, is the proper algebraical representative of the preceding series? If it have one, let us call it P; then, since the series itself is the same as

$$1+x\left\{1+x+x^2+ \text{\&c. } \textit{ad infinitum}\right\}$$

P cannot be used for the above unless $1+x\mathrm{P}$ may be substituted for P. Or we must have

$$\mathrm{P} = 1+x\mathrm{P} \text{ which gives } \mathrm{P} = \frac{1}{1-x}$$

the same result as when x is less than unity. And we may shew, in the same way as in the last page, that any algebraical operation performed either upon $\frac{1}{1-x}$, or upon $1+x+x^2+$ &c. gives but one result: indeed, since there is in algebraical multiplication no stipulation that the quantities employed must be positive, the process there employed is equally applicable in the case where x is greater than 1. But in this case we cannot obtain (as in last page) any approach to arithmetical equality, but the direct reverse; for x^{n+1} and x^{n+2} increase, instead of decreasing, as n increases.

The method of expanding algebraical quantities in pages 191, &c. does not require that x shall lie within the limits of convergency: but the process is equally conclusive in all cases. To expand $\frac{1}{1+x}$, for instance, we ask what expression of the form a_0+a_1x+ &c. will be the same as $\frac{1}{1+x}$. All we know of the latter lies in its definition, that, multiplied by $1+x$, it gives 1. The series $1-x+x^2-x^3+$ &c. has this property, independently of the value of x.

We shall now ascertain, 1st, that so far as instances can shew it, we may prove that this general use of the sign $=$ will not lead to inconsistent results: 2d, that we are not obliged to depend upon such a species of proof, but that the result is one which necessarily follows from the nature of our primary assumptions.

To try the first, let us *assume* that, in all cases,

$$\frac{1}{1+x} = 1-x+x^2-x^3+x^4+ \text{ \&c. } ad\ inf.$$

We have then, if $x = 1$,

$$\frac{1}{2} = 1-1+1-1+1- \text{ \&c. } ad\ inf.$$

a result which is in no sense the expression of an arithmetical equality; for the above series continued to an even number of terms must be 0, and to an odd number of terms, 1. We shall now try whether alternate addition and subtraction of a quantity to and from itself, *ad infinitum*, will or will not, when treated algebraically, give the half of the quantity in question. Let us suppose

$$\left.\begin{array}{rl} P = & 1+x+x^2+x^3+ \text{ \&c.} \\ -P = & -1-x-x^2- \text{ \&c.} \\ +P = & +1+x+ \text{ \&c.} \\ -P = & -1- \text{ \&c.} \end{array}\right\} (A)$$

$$\therefore\ P-P+P- \text{\&c.} = 1+(x-1)+x^2-x+1+ \text{ \&c.}$$

(page 103)

$$= \frac{x+1}{x+1} + \frac{x^2-1}{x+1} + \frac{x^3+1}{x+1} + \text{ \&c.}$$

$$= \frac{x+x^2+x^3+ \text{\&c.}}{x+1} + \frac{1-1+1- \text{\&c.}}{x+1}$$

But $P = \frac{1}{1-x}$, and $x+x^2+x^3+ \text{\&c.} = x(1+x+x^2+ \text{\&c.}) = \frac{x}{1-x}$; if, therefore, the preceding process be algebraically equivalent to halving $\frac{1}{1-x}$ and if $1-1+1-1+$ &c. may be changed into $\frac{1}{2}$, we have the following equation:

$$\frac{1}{2}\,\frac{1}{1-x} = \frac{\frac{x}{1-x}}{x+1} + \frac{1}{2}\,\frac{1}{x+1}$$

or

$$\frac{1}{2}\,\frac{1}{1-x} = \frac{x}{1-x^2} + \frac{1}{2}\,\frac{1}{x+1}$$

which will be found to be true.

The above contains the algebraical artifice of writing the series (A) in the form

$$\begin{array}{l} 1+x+x^2+x^3+ \text{ \&c.} \\ \quad -1-x-x^2- \text{ \&c.} \\ \qquad \text{\&c.} \qquad \text{\&c.} \end{array}$$

instead of

$$\begin{array}{c} 1+x+x^2+x^3+\text{ \&c.} \\ -1-x-x^2-x^3+\text{ \&c.} \\ \text{\&c.} \qquad \text{\&c.} \end{array}$$

or $$(1-1+1+\text{ \&c.})+(x-x+x+\text{ \&c.})+\text{\&c.}$$

But it is to be observed, that we do not say that every algebraical use of $=$ will produce arithmetical equalities, but only that whenever an algebraical use of $=$ *does* produce an arithmetical equation, we shall find that equation to be arithmetically true: an assertion as yet uncontradicted.

We now proceed to the second point.

We have previously so constructed the meaning of the fundamental symbols of algebra, that algebra, in certain cases, coincides entirely with arithmetic; and, more than this, the rules which follow from the definitions are so constructed, that when the result only is arithmetical, and preceded by algebraical steps, the alterations necessary to make these steps arithmetical, produce no alteration in the result. This being the case at the outset, and it being shewn that the number of steps through which we pass by algebraical process does not affect the preceding statement, we then know, 1st, that all arithmetical results so deduced may be depended upon, as much as if they were arithmetically deduced; 2d, that all results which are not explicable arithmetically, are such as are perfectly consistent with the definitions laid down; and, if not always arithmetically true, cannot produce a result which shall be arithmetical and false.

The reason why we have appealed to instances is, that the preceding argument, being general and abstruse, will not be thoroughly understood by the student until he has a degree of exact comprehension of the words employed, which can only be gained by familiarity with the use of algebraical language. There are also principles of reasoning,* *independent of algebra*, which are difficult,

* Remember that *mathematical reasoning* means *reasoning applied to mathematics*, and is not a different kind of reasoning from any other. The art of reasoning is *exercised* by mathematics, not *taught* by it; nor is the mathematician obliged to use one single principle which is not employed in every other branch of reasoning. In fact, an opposite of this is the case; there are principles in other branches of reasoning which are not employed in most branches of mathematics. Those who are not

and which the beginner will therefore not conquer by algebra alone. Such is the following: that if assertions which are not inconsistent with each other are rationally and logically used, the conclusions cannot be inconsistent with each other. But though this, in its full extent, be a very difficult proposition for a beginner to understand, the difficulty is not an algebraical one.

We have said, or at least implied, that when an arithmetical result is produced, the steps by which it was produced, if not already arithmetical, may be made so. Under arithmetical equality we include not only the absolute notion of equality which we recognise in $4+5=9$, but what we have called the *quam proximè* equality,* which we see in

$$\frac{1}{1-\frac{1}{2}} \text{ or } 2 = 1 + \frac{1}{2} + \frac{1}{4} + \frac{1}{8} + \text{\&c. } ad\ inf.$$

Let us now suppose that a is less than 1, but that we may suppose it as near to 1 as we please. Also, let x be less than 1 (which

aware of this talk of mathematical demonstration as if it were distinct from all other kinds.

But this is not the case; mathematical demonstration is, so far as it goes, the same thing as any other demonstration; the superior safety of mathematics lying, not in the method of arguing, but in the reasoner knowing more exactly what he is talking about than is usual in other discussions. The mathematical sciences are concerned in whatever can be counted or measured; and whoever talks of *mathematical* demonstration, as applied to any thing else, either means merely logical or certain demonstration, or does not know what he is talking about.

The opponents of mathematics are, of all men, those who pay this undue respect to *mathematical* reasoning; which they invariably do, by asserting as a discovery and a triumph, that those who are only mathematicians frequently reason ill on other subjects. We recommend the student to believe this whenever he meets with it, and to act accordingly; for it is an important truth, though not either a great or recent discovery—having been, in point of fact, ascertained immediately after the fall by our common ancestor, who, having till then been nothing but a gardener, must have found himself but an indifferent tailor.

* We are not here stepping even beyond the bounds of ordinary arithmetic. For $\sqrt{10}$, $\sqrt[3]{11}$, &c. have no other but a *quam proximè* existence; we can find fractions which, multiplied by themselves, shall be as near as we please to 10; we can sum 1, $\frac{1}{2}$, $\frac{1}{4}$, &c. until we come as near as we please to 2.

is necessary to the arithmetical existence of the final equation), and we have

$$\left.\begin{aligned}\frac{1}{1+a} &= 1 - a + a^2 - a^3 + \&c. \\ \frac{1}{1-x} &= 1 + x + x^2 + x^3 + \&c.\end{aligned}\right\} \text{arithmetically} \quad \begin{aligned}(1)\\(2)\end{aligned}$$

$$\begin{aligned}\frac{1}{1+a}\cdot\frac{1}{1-x} &= 1 + x + x^2 + x^3 + \&c. \\ &\quad - a - ax - ax^2 - \&c. \\ &\quad\quad + a^2 + a^2x + \&c. \\ &\quad\quad\quad - a^3 - \&c.\end{aligned}$$

$$= 1 - (a - x) + (x^2 - ax + a^2) - \&c. \text{ supposing } a > x$$

$$= \frac{a+x}{a+x} - \frac{a^2-x^2}{a+x} + \frac{a^3+x^3}{a+x} - \&c.$$

$$= \frac{a - a^2 + a^3 - \&c.}{a+x} + \frac{x + x^2 + x^3 + \&c.}{a+x}$$

$$= \frac{a}{1+a}\cdot\frac{1}{a+x} + \frac{x}{1-x}\cdot\frac{1}{a+x}$$

So long as we suppose a less than 1 (no matter how little), the preceding process is arithmetical; but the moment we suppose $a = 1$, the equation (1) loses all arithmetical character. But still, the last equation reduces itself to the one given in page 197.

We have chosen a very simple case, but we might, with operations of sufficient length, and by various artifices, reduce any algebraical equality to an arithmetical one. But methods of doing this, at once short and general, cannot yet be understood by the student.

Let x be first supposed positive; let it then become 0, and afterwards negative. The following and similar tables may then be made.

Sign of x	+	0	−
Sign of $\frac{1}{x}$	+	$\propto$	−
...... x^3	+	0	−
...... $\frac{1}{x^3}$	+	$\propto$	−
...... x^2	+	0	+
...... $\frac{1}{x^2}$	+	$\propto$	+

These, and other instances, give the following principle: *If, when* x *changes from* a *to* b, *passing through all intermediate values, the sign of a function of* x *change from positive to negative, or* vice versâ, *the point at which the change takes place is marked by its value being either* nothing *or* infinite; *but the converse is* not *true, that a function always changes its sign when its value becomes nothing or infinite.*

The *infinite* here spoken of is the form $\frac{a}{0}$; but, as we have seen, all methods of obtaining an arithmetical infinite (we should rather say all arithmetical methods of increasing number without limit), are not properly represented by $\frac{a}{0}$. If we take the equation

$$\frac{1}{1-x} = 1 + x + x^2 + x^3 + \text{ \&c. } ad\ inf.$$

we see that when x is greater than 1, the second side is, arithmetically speaking, only an indication of a method of obtaining number without limit, while the first side is negative. There is then a change of sign when x passes, say from $\frac{1}{2}$ to 2, and the change takes place when $x = 1$, giving,

$$2 = 1 + \frac{1}{2} + \frac{1}{4} + \frac{1}{8} + \text{ \&c.}$$

$$\frac{1}{0} = 1 + 1 + 1 + 1 + \text{ \&c.}$$

$$-\frac{1}{2} = 1 + 2 + 4 + 8 + \text{ \&c.}$$

We therefore see that a divergent series may be the algebraical representative of a negative quantity: but the series

$$\frac{1}{(1-x)^2} = 1 + 2x + 3x^2 + 4x^3 + \text{\&c.}$$

which is divergent when x is greater than 1, gives a positive result in all cases. From this we see that *when an equality specified is purely algebraical, we are not at liberty to compare magnitudes by any arithmetical comparison, if infinite series be in question.* For instance, if a be greater than a', b greater than b', &c., we may say that $a + b +$ &c. is greater than $a' + b' +$ &c.: 1st, so long as the number of both is finite; 2d, if a, b, &c. be so related that $a + b +$ &c. can never exceed a given limit. But we may not draw this conclusion whan $a+b+$ &c. increases without limit; nor may we say that the algebraical representative of $1+2+4+$ &c. is greater than that of $1+\frac{1}{2}+\frac{1}{4}+$ &c.

There is no liability to trangress this rule, because the quantity in question must cease to be the object of arithmetic before the reasoning upon it can fail. Let it then be observed, that *with regard to divergent series, we admit no results of comparison except those which are derived from their equivalent finite algebraical expressions.*

CHAPTER X.

ON THE NOTATION OF FUNCTIONS.

We have already defined what is meant by a *function* of x (see page 168): the present chapter is intended to exemplify the method of denoting a function, either for abbreviation, or because we wish to allot a specific symbol to some unknown function.

When we have to consider an expression, such as x^2+ax, only with reference to the manner in which it contains x, that is, to reason upon properties which depend upon the manner in which x enters, and the value of x, and not upon the manner of containing a, or its value, we call the expression a function of x, and denote it by a letter placed before x, in the manner of a coefficient. But to avoid confounding this functional symbol with that of a coefficient, certain letters are set apart always to denote functional symbols, and never coefficients. These letters will be, in the present work, F, f, ϕ, ψ. Thus, by $\mathrm{F}x$, fx, ϕx, ψx, we mean simply functions of x, given or not, as the case may be. By $\mathrm{F}\phi x$ we mean the same function of ϕx which $\mathrm{F}x$ is of x: for instance, if $\mathrm{F}x = x + x^2$, $\mathrm{F}\phi x$ is $\phi x + (\phi x)^2$.

EXAMPLES. Let $\phi x = 1+x^2 \quad \mathrm{F}x = 1-x^2$

$$\mathrm{F}\phi x = 1-(1+x^2)^2 = -2x^2-x^4$$

$$\phi \mathrm{F}x = 1+(1-x^2)^2 = 2-2x^2+x^4$$

Let $\phi x = x^a$ then $\phi(1+x) = (1+x)^a \quad \phi(2x) = (2x)^a$

$\phi(a) = a^a \qquad \phi b = b^a$ &c.

A *functional equation* is an equation which is necessarily true of a function or functions for every value of the letter which it contains. Thus, if $\phi x = ax$, we have $\phi(bx) = abx = b \times \phi x$, or

$$\phi(bx) = b\phi x$$

is always true when ϕx means ax.

Thus the following equations may be deduced:

If $\phi x = x^a \qquad \phi x \times \phi y = \phi(xy)$

$\phi x = a^x \qquad \phi x \times \phi y = \phi(x+y)$

$$\varphi x = ax + b \qquad \frac{\phi x - \phi y}{\phi x - \phi z} = \frac{x - y}{x - z}$$

$$\varphi x = ax \qquad \varphi x + \varphi y = \varphi(x+y)$$

We can often, from a functional equation, deduce the algebraical form which will satisfy it: for instance, if we know that $\phi(xy) = x \times \phi y$, supposing this always true, it is true when $y = 1$, which gives $\phi(x) = x \times \phi(1)$. But $\phi(1)$ is an independent quantity, made by writing 1 instead of x in $\phi(x)$. Let us call it c: it only remains to ascertain whether any value of c will satisfy the equation. Let $\phi x = cx$; then $\phi(xy) = cxy$, and $x \times \phi y = x \times cy = cxy$; whence $\phi(xy) = x\phi y$ for all values of c, and ϕx being cx, $\phi(1)$ is $c \times 1$ or c, as was supposed. Similarly, if

$\varphi(xy) = (\varphi x)^y$, by making $x = 1$ we have

$$\varphi y = \{ \varphi(1) \}^y = c^y \qquad \varphi x = c^x$$

$$\varphi(xy) = c^{xy} = (c^x)^y = (\varphi x)^y$$

and $\phi(1) = c^1 = c$, as was supposed. This, with the following theorem, will be sufficient for the investigations connected with functions which we shall need in this work.

We have seen that if $\phi x = c^x$ we have $\phi x \times \phi y = \phi(x+y)$; but we do not know whether there may not be other functions of x which have the same property. We shall now, however, prove this converse, namely, that the equation $\phi x \times \phi y = \phi(x+y)$ can be satisfied by no other function of x, except those of the form c^x.

Let us suppose ϕx to be a function of such a character that *whatever may be the values of* x *and* y, the equation

$$\varphi x \times \varphi y = \varphi(x+y) \cdots\cdots (A)$$

is true. For y write $a+b$, which gives

$$\varphi x \times \varphi(a+b) = \varphi(x+a+b)$$

But equation (A), as described, gives $\phi(a+b) = \phi a \times \phi b$, whence

$$\varphi x \times \varphi a \times \varphi b = \varphi(x+a+b)$$

For either letter, a, for instance, write $c+e$, which gives

$$\varphi x \times \varphi(c+e) \times \varphi b = \varphi(x+c+e+b)$$

but

$$\varphi(c+e) = \varphi c \times \varphi e$$

whence

$$\varphi x \times \varphi c \times \varphi e \times \varphi b = \varphi(x+c+e+b)$$

and so on: that is, if there be n quantities of any value whatsoever, namely, a_1, a_2, a_3 ...... a_{n-1}, a_n, we have

$$\phi a_1 \times \phi a_2 \cdots \times \phi a_{n-1} \times \phi a_n = \phi(a_1 + a_2 + \cdots + a_{n-1} + a_n)$$

Let these n quantities be all equal to each other and to a, which gives

$$\left\{\begin{matrix}\phi a \times \phi a \times \cdots\cdots \times \phi a \times \phi a \\ \text{there being } n \text{ factors.}\end{matrix}\right\} = \phi\left(\begin{matrix}a + a + \cdots\cdots + a + a \\ \text{there being } n \text{ terms.}\end{matrix}\right)$$

or $$(\phi a)^n = \phi(na)$$

where n is any whole number.*

Similarly, if we had supposed m quantities each equal to b, we should have had

$$(\phi b)^m = \phi(mb)$$

Now, m and n being whole numbers, let us suppose $mb = na$, which gives †

$$\phi(mb) = \phi(na) \quad \text{or} \quad (\phi b)^m = (\phi a)^n$$

or $$\phi b = (\phi a)^{\frac{n}{m}} \quad \text{but} \quad b = \frac{n}{m}a, \text{ whence}$$

$$\phi\left(\frac{n}{m}a\right) = (\phi a)^{\frac{n}{m}}$$

or the equation $\phi(pa) = (\phi a)^p$ is true when p is a whole number, or a commensurable fraction (page 98).

In the original equation (A), let $x = 0$ and $y = 0$, whence $x + y = 0$. Let $\phi(0)$ be called c; we have then $c \times c = c$, or $c = 1$. Then let $y = -x$, or $x + y = 0$, which gives

$$\phi x \times \phi(-x) = \phi(0) = 1 \quad \text{or} \quad \phi(-x) = \frac{1}{\phi x};$$

which being true for every value of x, is true if for x we write pa, p being a whole number or commensurable fraction. This gives

$$\phi(-pa) = \frac{1}{\phi(pa)} = \frac{1}{(\phi a)^p} = (\phi a)^{-p}$$

* We cannot conceive a fraction of the preceding process. (See page 37).

† Similar operations performed upon equal quantities must give equal results.

or the equation

$$\varphi(pa) = (\varphi a)^p$$

is true if p be a negative whole number or commensurable fraction. Hence, as in page 204, by making $a = 1$, we determine

$$\varphi(p) = c^p$$

where (provided p be commensurable) c may be any quantity whatever.

The preceding equation will also be true when p is an incommensurable quantity, such as $\sqrt{2}$, $\sqrt[3]{4}$, &c.; but the method by which we shall treat the consequences of this equation will render a proof unnecessary.

Exercise. Shew that the equation

$$\varphi(x+y) + \varphi(x-y) = 2\varphi x \times \varphi y$$

is satisfied by

$$\varphi x = \frac{1}{2}\left(a^x + a^{-x}\right)$$

for every value of a: and also that

$$\varphi(x+y) = \varphi x + \varphi y$$

can have no other solution than

$$\varphi x = ax$$

CHAPTER XI.

ON THE BINOMIAL THEOREM.

THE binomial theorem is a name given to the method of expanding $(a+b)^n$ into a series (finite or infinite as the case may be) of powers of a and b, n being either whole or fractional, positive or negative commensurable or incommensurable.

The preceding case may be reduced to that of expanding $(1+x)^n$ in a series of powers of x: for

$$a+b=a\left(1+\frac{b}{a}\right) \qquad \left(a+b\right)^n=a^n\left(1+\frac{b}{a}\right)^n$$

Let $x=\frac{b}{a}$, and $(1+x)^n$ is the function to be expanded.

As this theorem is of particular importance, we shall give two investigations of it: the first analytical, inquiring what is the series which is equivalent to $(1+x)^n$; the second synthetical, shewing that the series so found is the one required.

If it be possible, let $(1+x)^n$ be a series of whole powers of x of the form

$$(1+x)^n=a_0+a_1x+a_2x^2+a_3x^3+\&c.$$

in which a_0, a_1, &c. are functions of n and not of x.

LEMMA. Whatever may be the value of n, the limit of $\frac{a^n-b^n}{a-b}$, to which it approaches as b is made more and more nearly equal to a, is na^{n-1}.

[When $a=b$, observe that this fraction takes the form $\frac{0}{0}$, page 162.]

First, let n be a whole number. Then, page 103,

$$\frac{a^n-b^n}{a-b}=a^{n-1}+a^{n-2}b+a^{n-3}b^2+\ \cdots\ +ab^{n-2}+b^{n-1}$$

as b and a approach to equality, the limit of the second side is

$$a^{n-1}+a^{n-2}a+a^{n-3}a^2+\ \cdots\ +a\,a^{n-2}+a^{n-1}$$

or $\quad a^{n-1} + a^{n-1} + a^{n-1} + \cdots + a^{n-1} + a^{n-1}$

or $\quad n a^{n-1}$

[That there are n terms in the preceding is evident from this, that there is a term for every power of b from 1 to $n-1$, both inclusive, and one term independent of b.]

Secondly, let n be a fraction $\frac{p}{q}$, where p and q are whole numbers. We have then

$$\frac{a^n - b^n}{a-b} = \frac{a^{\frac{p}{q}} - b^{\frac{p}{q}}}{a-b} = \frac{\left(a^{\frac{1}{q}}\right)^p - \left(b^{\frac{1}{q}}\right)^p}{\left(a^{\frac{1}{q}}\right)^q - \left(b^{\frac{1}{q}}\right)^q}$$

$$(\text{Let } a^{\frac{1}{q}} = a_{,} \qquad b^{\frac{1}{q}} = b_{,}) \quad = \frac{a_{,}^p - b_{,}^p}{a_{,}^q - b_{,}^q}$$

$$= \frac{\dfrac{a_{,}^p - b_{,}^p}{a_{,} - b_{,}}}{\dfrac{a_{,}^q - b_{,}^q}{a_{,} - b_{,}}}$$

now, as a approaches to b, $a_{,}$ approaches to $b_{,}$, and as p and q are whole numbers, the limits of the numerator and denominator of the preceding are $p a_{,}^{p-1}$ and $q a_{,}^{q-1}$; whence the limit of the fraction i

$$\frac{p a_{,}^{p-1}}{q a_{,}^{q-1}} \text{ or } \frac{p}{q} a_{,}^{p-q} \text{ or } \frac{p}{q}\left(a^{\frac{1}{q}}\right)^{p-q} \text{ or } \frac{p}{q} a^{\frac{p}{q}-1} \text{ or } n a^{n-1}$$

Thirdly, let n be negative, and let the corresponding positive quantity be p, so that $n = -p$. Then

$$\frac{a^n - b^n}{a-b} = \frac{a^{-p} - b^{-p}}{a-b} = \frac{\dfrac{1}{a^p} - \dfrac{1}{b^p}}{a-b} = \frac{1}{a^p b^p} \cdot \frac{b^p - a^p}{a-b}$$

$$= -\frac{1}{a^p b^p} \times \frac{a^p - b^p}{a-b}$$

as a approaches towards b, the limit of the first factor is $-\frac{1}{a^p a^p}$ or $-a^{-2p}$, and that of the second (p being positive) is, as already proved, $p a^{p-1}$. Hence the limit of the preceding product is $-a^{-2p} \times p a^{p-1}$ or $-p a^{-p-1}$, which, since $n = -p$, is $n a^{n-1}$.

We now resume the assumed series

$$(1 + x)^n = a_0 + a_1 x + a_2 x^2 + a_3 x^3 + \&c.$$

Let y be a quantity which may be made as near to x as we please; the above series being supposed true for all values of x, we may put y in the place of x, which gives

$$(1+y)^n = a_0 + a_1 y + a_2 y^2 + \&c.$$

$$(1+x)^n - (1+y)^n = a_1(x-y) + a_2(x^2-y^2) + \&c.$$

Divide both sides by $x-y$, or $(1+x)-(1+y)$.

$$\frac{(1+x)^n-(1+y)^n}{(1+x)-(1+y)} = a_1 + a_2\frac{x^2-y^2}{x-y} + a_3\frac{x^3-y^3}{x-y} + \&c.$$

which two sides being always equal, the limits to which they approach, as x and y approach to equality (in which case $1+x$ and $1+y$ approach to equality), are equal; or

$$n(1+x)^{n-1} = a_1 + 2a_2 x + 3a_3 x^2 + \&c.$$

Multiply both sides by $1+x$.

$$n(1+x)^n = a_1 + 2a_2x + 3a_3x^2 + 4a_4x^3 + \&c.$$
$$+ \quad a_1x + 2a_2x^2 + 3a_3x^3 + \&c.$$

But

$$n(1+x)^n = na_0 + na_1x + na_2x^2 + na_3x^3 + \&c.$$

therefore, page 188,

$$a_1 = na_0, \quad 2a_2 + a_1 = na_1 \text{ or } a_2 = \frac{n-1}{2}a_1 = n\frac{n-1}{2}a_0$$

$$3a_3 + 2a_2 = na_2 \quad \text{or} \quad a_3 = \frac{n-2}{3}a_2 = n\,.\frac{n-1}{2}\,.\frac{n-2}{3}a_0$$

$$4a_4 + 3a_3 = na_3 \quad \text{or} \quad a_4 = \frac{n-3}{4}a_3 = n\frac{n-1}{2}\frac{n-2}{3}\frac{n-3}{4}a_0$$

from which, by substitution and observing that a_0 is a factor of all the terms, we have

$$(1+x)^n = a_0\left(1 + nx + n\frac{n-1}{2}x^2 + n\frac{n-1}{2}\frac{n-2}{3}x^3 + \&c.\right)$$

In which a_0 is not yet determined. To determine it, we must first see whether the preceding series is ever convergent.

But previously, we must observe that what we have proved is not, properly speaking, the truth of the above equation, but only that, *if* $(1+x)^n$ *can always be expanded in a series of whole powers of* x, *the preceding series is the one.* For we have assumed that $(1+x)^n$ is $a_0 + a_1x + a_2x^2 + \&c.$, when, for any thing we know to the contrary, it might be a series of fractional, or negative, or mixed powers which should have been chosen. Whenever, in our previous investigations, we have made an impossible supposition, we have always been warned of it at the end of the process by the appearance of some new and unexplained anomaly. As we have not, before this one, made any assumption of the form of an expansion other than we had either reason or experience to justify, and as we have neither for the one we have actually made, we know neither whether we are right, nor what is the index of our being wrong. It may be that the appearance of an always-divergent series would follow from a mistake, if there be any. We therefore examine the preceding series.

The ratios of the several terms to those immediately preceding are

$$nx \qquad \frac{n-1}{2}x \qquad \frac{n-2}{3}x \qquad \frac{n-3}{4}x \quad \&c.$$

the general form of which is

$$\frac{(p+2)\text{ term}}{(p+1)\text{ term}} = \frac{n-p}{p+1}x$$

First, we observe, that if n be a positive whole number, the series is finite: for the $(n+2)$th term will contain $n-n$ or 0 as a factor, and so will all the following terms. We therefore take the case where n is fractional or negative. When p has passed n, the preceding ratio will be always negative, shewing that the terms become alternately positive and negative; for that the ratio of two quantities is negative indicates that they differ in sign. Neglect the sign of the preceding, and make it positive. This we may do, as our object is to see whether the series can be made convergent, which a series of terms alternately positive and negative always can be, if the corresponding series of positive terms can be made convergent. We have then

$$\frac{p-n}{p+1}x \quad \text{or} \quad \frac{px}{1+p} - \frac{nx}{1+p} \quad \text{or} \quad \frac{x}{1+\frac{1}{p}} - \frac{nx}{1+p}$$

as we take higher and higher terms, the second term of the preceding diminishes without limit, and the first has the limit x. Consequently, if x be less than 1, the ratio above mentioned will, after a certain number of terms, become less than unity, and will afterwards approximate continually to the limit x. That is, the series obtained is always convergent when x is less than 1.

This being the case, we may, page 189, use the result of making

$x=0$, which gives $(1)^n = a_0$. If n be a whole number, $a_0 = 1$; but if n be a fraction, as $\frac{p}{q}$ we have $(1)^{\frac{p}{q}} = a_0 = (1^p)^{\frac{1}{q}} = (1)^{\frac{1}{q}}$, or a_0 may be any qth root of 1; see page 113. If we choose the arithmetical root, we find $a_0 = 1$; and this series is (if the doubtful assumption be true) the arithmetical nth power of $1+x$, when x is less than 1, or the algebraical equivalent of $(1+x)^n$ in all cases.

We shall now shew that the series is correct whenever n is a whole number. Suppose it correct for any one whole number, say m. Then (a_0 being 1),

$$(1+x)^m = 1 + mx + m\frac{m-1}{2}x^2 + m\frac{m-1}{2}\frac{m-2}{3}x^3 + \text{ \&c.}$$

Multiply both sides by $1+x$.

$$(1+x)^{m+1} = 1 + mx + m\frac{m-1}{2}x^2 + m\frac{m-1}{2}\frac{m-2}{3}x^3 + \text{ \&c.}$$

$$+ \quad x \quad + m \qquad x^2 + m\frac{m-1}{2} \qquad x^3 + \text{ \&c.}$$

$$= 1 + (m+1)x + \left(m\frac{m-1}{2} + m\right)x^2 + \left(m\frac{m-1}{2}\frac{m-2}{3} + m\frac{m-1}{2}\right)x^3 + \text{ \&c.}$$

But $m\frac{m-1}{2} + m = m\left(\frac{m-1}{2} + 1\right) = m\frac{m+1}{2} = (m+1)\frac{m}{2}$

$$m\frac{m-1}{2}\frac{m-2}{3} + m\frac{m-1}{2} = m\frac{m-1}{2}\left(\frac{m-2}{3} + 1\right) = (m+1)\frac{m}{2}\frac{m-1}{3}$$

whence

$$(1+x)^{m+1} = 1 + (m+1)x + (m+1)\frac{m}{2}x^2 + (m+1)\frac{m}{2}\frac{m-1}{3}x^3 + \text{ \&c.}$$

which, if we now write n for $m+1$, or $n-1$ for m, becomes the same series, or follows the same law as

$$(1+x)^n = 1 + nx + n\frac{n-1}{2}x^2 + n.\frac{n-1}{2}\frac{n-2}{3}x^3 + \text{ \&c.}$$

Hence, if this expression be true for any one whole value of n, it is true for the next. But it is true when $n=1$; for

$$(1+x)^1 = 1 + 1x + 1.\frac{1-1}{2}x^2 + 1.\frac{1-1}{2}.\frac{1-2}{3}x^3 + \text{ \&c.}$$

therefore it is true when $n=2$; but it is therefore true when $n=3$, and so on, *ad infinitum*.

If we consider $(1+x)^n$ as a function of n, and call it ϕn, we see that

$$(1+x)^n \times (1+x)^m = (1+x)^{n+m}$$

or

$$\varphi n \quad \times \quad \varphi m \quad = \varphi(n+m)$$

but when n is a whole number $(1+x)^n$ is the series in question; therefore, calling the above series ϕn, we have, *when* n *is a whole number*,

$$\varphi n \times \varphi m = \varphi(n+m)$$

or

$$\left(1+nx+n.\frac{n-1}{2}x^2+n.\frac{n-1}{2}.\frac{n-2}{3}x^3+\&c.\right)$$

$$\times\left(1+mx+m\frac{m-1}{2}x^2+m\frac{m-1}{2}\frac{m-2}{3}x^3+\&c.\right)$$

$$=1+(m+n)x+(m+n)\frac{m+n-1}{2}x^2+(m+n)\frac{m+n-1}{2}\frac{m+n-2}{3}x^3+\&c.$$

This may be verified to any extent we please, by actual multiplication; for the two first series multiplied together give

$$1+(m+n)x+\left(n.\frac{n-1}{2}+nm+m\frac{m-1}{2}\right)x^2$$

$$+\left(n.\frac{n-1}{2}\frac{n-2}{3}+n.\frac{n-1}{2}m+nm.\frac{m-1}{2}+m.\frac{m-1}{2}\frac{m-2}{3}\right)x^3+\&c.$$

But $n.\frac{n-1}{2}+nm+m.\frac{m-1}{2}=\frac{n^2-n+2nm+m^2-m}{2}$

$$=\frac{(n+m)^2-(n+m)}{2}=(n+m)\frac{n+m-1}{2}$$

$$n.\frac{n-1}{2}\frac{n-2}{3}+n\frac{n-1}{2}m+nm\frac{m-1}{2}+m\frac{m-1}{2}\frac{m-2}{3}$$

$$=\frac{n^3-3n^2+2n+3n^2m-3nm+3nm^2-3nm+m^3-3m^2+2m}{2\times 3}$$

$$=\frac{(n+m)^3-3(n+m)^2+2(n+m)}{2\times 3}=(n+m)\frac{n+m-1}{2}.\frac{n+m-2}{3}$$

and so on. We now lay down the following principle: *When an algebraical multiplication, or other operation, such as has hitherto been defined, can be proved to produce a certain result in cases where the letters stand for whole numbers, then the same result must be true when the letters stand for fractions, or incommensurable numbers, and also when they are negative.* For we have never yet had occasion to distinguish results into those which are true for whole numbers, and those which are not true for whole numbers; but all processes have been, as stated in the introduction, true whether the letters are whole numbers or fractions. There has been no such thing in any process as a term of an equation, which exists when a letter stands for a whole number, but does not exist when it stands for a fraction. If, therefore, $\phi m \times \phi n$ will give $\phi(m+n)$ by the common process of multiplication when m and n are whole numbers, that same process will also give $\phi(m+n)$ when m and n are fractions, and also when they are one or both negative: consequently, the series

$$1+nx+n\frac{n-1}{2}x^2+ \text{ \&c.}$$

has this property, that, considered as a function of n, and called ϕn, it satisfies

$$\phi n \times \phi m = \phi(m+n)$$

in all cases. But in page 205, it has been proved that any solution of the preceding equation must be $\phi n = c^n$ where $c = \phi(1)$, and $\phi(1)$ we find to be

$$1+1x+1\frac{1-1}{2}x^2+ \text{ \&c. or } 1+x$$

therefore ϕn is in all cases $(1+x)^n$, which is the theorem in question.*

* Every proof which has ever been given of this theorem has been contested; that is, no one has ever disputed the truth of the theorem itself, but only the method of establishing it. And the general practice is, for each proposer of a new proof to be very much astonished at the want of logic of his predecessors. The proof given in the text is a combination of two proofs, the first part, making use of limits, given (according to Lacroix) in the *Phil. Trans.* for 1796; the second, the well-known proof of Euler. The objection to the first part lies in the assumption of a series of whole powers; to the second, in its being *synthetical*, that is, not finding what $(1+x)^n$ is, but only proving that a certain given series is the same as $(1+x)^n$. But each part of this proof answers the objection made to *the other part*; in the first part analysis is employed, but only so as to give strong grounds of conjecture that $1+nx+$ &c. is the required series; in the second part this conjectural (not arbitrarily chosen) series is absolutely shewn to be that required. The proof of Euler may be condensed into the following, of which the several assertions are proved in the text.

1. If $1+nx+n\,.\frac{n-1}{2}x^2+$ &c. be called ϕn, then $\phi(1)$ is $1+x$, and $\phi n \times \phi m$ is found to be $\phi(n+m)$.

2. If $\phi n \times \phi m = \phi(n+m)$ then ϕn must be $\{\phi(1)\}^n$

3. Therefore, $1+nx+$ &c. is $(1+x)^n$

In 1827, Messrs. Swinburne and Tylecote published their demon-

To use the preceding series, the readiest way is to form the several factors n, $\frac{n-1}{2}$, $\frac{n-2}{3}$, &c. previously to proceeding further; for instance, let $n=\frac{1}{2}$, or let it be $\sqrt{1+x}$, which is to be expanded.

$$n=\frac{1}{2},\ \frac{n-1}{2}=-\frac{1}{4},\ \frac{n-2}{3}=-\frac{1}{2},\ \frac{n-3}{4}=-\frac{5}{8}\ \text{\&c.}$$

$$\sqrt{1+x}=1+\frac{1}{2}x+\left(\frac{1}{2}\right)\left(-\frac{1}{4}\right)x^2+\left(\frac{1}{2}\right)\left(-\frac{1}{4}\right)\left(-\frac{1}{2}\right)x^3$$
$$+\left(\frac{1}{2}\right)\left(-\frac{1}{4}\right)\left(-\frac{1}{2}\right)\left(-\frac{5}{8}\right)x^4+\ \text{\&c.}$$
$$=1+\frac{1}{2}x-\frac{1}{8}x^2+\frac{1}{16}x^3-\frac{5}{128}x^4+\ \text{\&c.}$$

If $n=-1$,

$$n=-1,\ \frac{n-1}{2}=-1,\ \frac{n-2}{3}=-1,\ \frac{n-3}{4}=-1\ \text{\&c.}$$

$$(1+x)^{-1}\ \text{or}\ \frac{1}{1+x}=1+(-1)x+(-1)(-1)x^2$$
$$+(-1)(-1)(-1)x^3+\ \text{\&c.}$$
$$=1-x+x^2-x^3+\ \text{\&c.}$$

as has been proved before.

---

stration of this theorem, the last original one with which I am acquainted. It is much too laborious and difficult for a beginner, but is as unobjectionable in point of logic as I conceive the one given in the text to be. Their general objections to the theory of limits are not, I conceive, to its logical soundness, but to its applicability in algebra; it being more frequently than not a sort of convention that limits shall not be introduced into algebra. But, until $\sqrt{10}$ is explained in arithmetic without limits, I shall hold that limits are, and always have been, introduced into algebra. Nay, until the symbol 0 is well explained in all its uses, without limits, the same may be said. With regard to the objections made by the above-named gentlemen to Euler's proof, in their page 35, they evidently hang upon the assumption of a finite number of terms of the series, which is *not* Euler's supposition, and which he therefore "keeps out of sight," not "dexterously shuffles out of the way," for it does not come in the way, any more than lines which meet come in the way when we are expressly talking of parallel lines. After quoting the last-mentioned phrase, it is but fair to those gentlemen to state, that, judging by the general tenor of their work, it is not to be presumed that they meant to accuse Euler of intentional deceit, which the phrase "dexterous shuffling" generally means, and which, therefore, was not the proper phrase to use.

If $n = 5$,

$$n = 5 \quad \frac{n-1}{2} = 2 \quad \frac{n-2}{3} = 1 \quad \frac{n-3}{4} = \frac{1}{2} \quad \frac{n-4}{5} = \frac{1}{5} \quad \frac{n-5}{5} = 0$$

$$(1+x)^5 = 1+5x+\left|\begin{array}{l}5\\ \times 2\end{array}\right| x^2 + \left|\begin{array}{l}5\\ \times 2\\ \times 1\end{array}\right| x^3 + \left|\begin{array}{l}5\\ \times 2\\ \times 1\\ \times \frac{1}{2}\end{array}\right| x^4 + \left|\begin{array}{l}5\\ \times 2\\ \times 1\\ \times \frac{1}{2}\\ \times \frac{1}{5}\end{array}\right| x^5 + \left|\begin{array}{l}5\\ \times 2\\ \times 1\\ \times \frac{1}{2}\\ \times \frac{1}{5}\\ \times 0\end{array}\right| x^6 + \text{\&c.}$$

$$= 1+5x+10x^2+10x^3+5x^4+x^5+0+0+ \text{\&c.}$$

It is usual to prove the binomial theorem in the case of a whole exponent, as follows: Multiply $x+a$ by $x+b$, which gives $x^2+(a+b)x+ab$, and this again by $x+c$, which gives

$$x^3+(a+b+c)x^2+(ab+bc+ca)x+abc$$

From which it appears, that if there be n quantities, $a_1\, a_2 \ldots\ldots a_n$, and if the sum of all be called P_1, the sum of the products of every two, P_2 &c., that is, if

$$\begin{array}{lll}
P_1 & = & a_1 + a_2 + a_3 + \quad \ldots\ldots\ldots\ldots \\
P_2 & = & a_1 a_2 + a_2 a_3 + a_1 a_3 + \quad \ldots\ldots \\
P_3 & = & a_1 a_2 a_3 + a_1 a_3 a_4 + \quad \ldots\ldots\ldots\ldots \\
\ldots & \ldots & \ldots\ldots\ldots\ldots\ldots\ldots\ldots\ldots \\
\ldots & \ldots & \ldots\ldots\ldots\ldots\ldots\ldots\ldots\ldots \\
P_{n-1} & = & \left\{\begin{array}{c}\text{product of all}\\ \text{except } a_1\end{array}\right\} + \left\{\begin{array}{c}\text{product of all}\\ \text{except } a_2\end{array}\right\} + \ldots \\
P_n & = & \text{product of all;}
\end{array}$$

it follows that

$$(x+a_1)(x+a_2)\,..\,(x+a_n) = x^n + P_1 x^{n-1} + P_2 x^{n-2} + \,..\, + P_{n-1} x + P_n$$

The number of terms in P_1 is n; in P_2 as many as there are combinations of 2 out of n quantities, or (Ar. 210) $n \cdot \frac{n-1}{2}$; in P_3 as many as there are combinations of 3 out of n quantities, or $n\,\frac{n-1}{2}\,\frac{n-2}{3}$; and so on. Hence, if $a_1\ a_2$ &c. be all supposed equal to each other and to a, we have

$$P_1 = a + a + a + \cdots\cdots = na$$
$$P_2 = a^2 + a^2 + a^2 + \cdots\cdots = n\frac{n-1}{2}a^2$$
$$P_3 = a^3 + a^3 + a^3 + \cdots\cdots = n\frac{n-1}{2}\frac{n-2}{3}a^3$$

....

or $\left\{\begin{array}{c}(x+a)(x+a)\ldots\ldots(x+a) \\ \text{containing } n \text{ factors.}\end{array}\right\} = x^n + nax^{n-1} + n\frac{n-1}{2}a^2x^{n-2} + \&c.$

In which, if we suppose $x = 1$, we have

$$(1+a)^n = 1 + na + n\frac{n-1}{2}a^2 + \&c.$$

See Ar. 211, for the reason why the coefficients are the same, whether we begin from the one end or the other of the series, as will also appear in the following cases.

$$(1+x)^2 = 1+2x+x^2$$
$$(1+x)^3 = 1+3x+3x^2+x^3$$
$$(1+x)^4 = 1+4x+6x^2+4x^3+x^4$$
$$(1+x)^5 = 1+5x+10x^2+10x^3+5x^4+x^5$$
$$(1+x)^6 = 1+6x+15x^2+20x^3+15x^4+6x^5+x^6$$

In order to find $(1-x)^n$, change the sign of x in the series; that is, write $-x$ for x, leave x^2 unaltered; write $-x^3$ for x^3, and so on; which gives

$$(1-x)^n = 1 - nx + n\frac{n-1}{2}x^2 - \&c.$$

We may write the general series in the following way *when* n *is a whole number*, where P_n signifies the product of all whole numbers between 1 and n, both inclusive.

$$(1+x)^n = P_n\left\{\frac{1}{P_n} + \frac{x}{P_1 P_{n-1}} + \frac{x^2}{P_2 P_{n-2}} + \cdots\cdots + \frac{x^{n-1}}{P_{n-1}P_1} + \frac{x^n}{P_n}\right\}$$

which shews the similarity of the coefficients above alluded to.

We give the following as exercises:

1. When n is a whole number,

$$2^n = 1 + n + n\frac{n-1}{2} + n\frac{n-1}{2}\frac{n-2}{3} + \&c.$$

$$0 = 1 - n + n\frac{n-1}{2} - n\frac{n-1}{2}\frac{n-2}{3} + \&c.$$

$$2^{n-1} = 1 + n\frac{n-1}{2} + n\frac{n-1}{2}\,\frac{n-2}{3}\,\frac{n-3}{4} + \text{ \&c.}$$

2. $(a+b)^n = a^n + n a^{n-1} b + n\frac{n-1}{2} a^{n-2} b^2 + \text{ \&c.}$

3. If n be a whole number

$$\left(x+\frac{1}{x}\right)^2 = x^2 + \frac{1}{x^2} + 2$$

$$\left(x+\frac{1}{x}\right)^3 = x^3 + \frac{1}{x^3} + 3\left(x+\frac{1}{x}\right)$$

$$\left(x+\frac{1}{x}\right)^4 = x^4 + \frac{1}{x^4} + 4\left(x^2+\frac{1}{x^2}\right) + 6$$

$$\left(x+\frac{1}{x}\right)^{2n} = x^{2n} + \frac{1}{x^{2n}} + 2n\left(x^{2n-2} + \frac{1}{x^{2n-2}}\right)$$

$$+ 2n\,\frac{2n-1}{2}\left(x^{2n-4} + \frac{1}{x^{2n-4}}\right) + \ldots\ldots$$

ending with $\dfrac{2n(2n-1).(2n-2)\ \ldots\ldots\ (n+1)}{1\ .\ 2\ .\ 3\ .\ \ldots\ldots\ n}$

$$\left(x+\frac{1}{x}\right)^{2n+1} = x^{2n+1} + \frac{1}{x^{2n+1}} + (2n+1)\left(x^{2n-1} + \frac{1}{x^{2n-1}}\right) + \ldots\ldots$$

ending with $\dfrac{(2n+1)\,(2n)\ \ldots\ldots\ (n+2)}{1\ .\ 2\ .\ 3\ \ \ldots\ldots\ (n+1)}\left(x+\frac{1}{x}\right)$

4. $\dfrac{(1+x)^n + (1-x)^n}{2} = 1 + n\dfrac{n-1}{2}x^2 + n\dfrac{n-1}{2}\,\dfrac{n-2}{3}\,\dfrac{n-3}{4}x^4 + \text{ \&c.}$

$$\frac{(1+x)^n - (1-x)^n}{2} = nx + n\frac{n-1}{2}\,\frac{n-2}{3}x^3 + \text{ \&c.}$$

5. The student may provide himself with examples and verifications in the following way: choose any exponent n, whole or fractional, positive or negative, and expand $(1+x)^n$ and $(1+x)^{n+1}$ by means of the theorem; then multiply the first series so found by $1+x$, which should give the second series.

6. $a^n = b^n + n(a-b)b^{n-1} + n\dfrac{n-1}{2}(a-b)^2 b^{n-2} + \text{ \&c.}$

The following cases will afterwards require particular notice:

$$\left(1+\frac{1}{n}\right)^{nx} = 1 + nx\frac{1}{n} + nx\frac{nx-1}{2}\,\frac{1}{n^2} + nx\,\frac{nx-1}{2}\,\frac{nx-2}{3}\,\frac{1}{n^3} + \text{ \&c.}$$

But $\quad nx\dfrac{1}{n} = x \qquad nx\dfrac{nx-1}{2} \times \dfrac{1}{n^2} = x\,\dfrac{x-\frac{1}{n}}{2}$

$$nx\,\frac{nx-1}{2}\,\frac{nx-2}{3}\,\frac{1}{n^3} = x\,\frac{x-\frac{1}{n}}{2}\,\frac{x-\frac{2}{n}}{3} \quad \text{\&c. \&c. whence}$$

$$\left(1+\frac{1}{n}\right)^{nx} = 1 + x + x\frac{x-\frac{1}{n}}{2} + x\frac{x-\frac{1}{n}}{2}\frac{x-\frac{2}{n}}{3} + \&c.$$

In the preceding let $x = 1$, which gives

$$\left(1+\frac{1}{n}\right)^{n} = 1 + 1 + \frac{1-\frac{1}{n}}{2} + \frac{1-\frac{1}{n}}{2}\frac{1-\frac{2}{n}}{3} + \&c.$$

But (page 105), $\left(1+\frac{1}{n}\right)^{nx} = \left\{\left(1+\frac{1}{n}\right)^{n}\right\}^{x}$

or the first series is the xth power of the second, and

$$\left(1 + 1 + \frac{1-\frac{1}{n}}{2} + \&c.\right)^{x} = 1 + x + x\frac{x-\frac{1}{n}}{2} + \&c.$$

If $n = 0$, $(1+x)^n = 1$, or $(1+x)^n - 1 = 0$, or the fraction $\frac{(1+x)^n-1}{n}$ assumes the form $\frac{0}{0}$. We now ask whether this fraction has a limit when n is diminished without limit (page 162).

From the general theorem,

$$\frac{(1+x)^n-1}{n} = x + \frac{n-1}{2}x^2 + \frac{n-1}{2}\cdot\frac{n-2}{3}x^3 + \&c.$$

and when n is diminished without limit, the limit of the second side is

$$x + \left(\frac{-1}{2}\right)x^2 + \left(\frac{-1}{2}\right)\left(\frac{-2}{3}\right)x^3 + \left(\frac{-1}{2}\right)\left(\frac{-2}{3}\right)\left(\frac{-3}{4}\right)x^4 + \&c.$$

or $$x - \frac{x^2}{2} + \frac{x^3}{3} - \frac{x^4}{4} + \&c.$$

That is,

As n diminishes without limit, $\frac{(1+x)^n-1}{n}$ approaches without limit to . . . . $\Big\}$ $x - \frac{x^2}{2} + \frac{x^3}{3} - \frac{x^4}{4} + \&c.$

If $x = z - 1$, then the limit of $\frac{z^n-1}{n}$ (n diminishing without limit) is

$$(z-1) - \frac{1}{2}(z-1)^2 + \frac{1}{3}(z-1)^3 - \&c.$$

Of this series we know no property as yet, except that it is convergent when x or $z-1$ is less than 1, or z less than 2. We shall proceed to examine the expression from which it is derived. If in it we write z^m for z, we have

$$\frac{z^{mn}-1}{n} \quad \text{or} \quad m\frac{z^{mn}-1}{mn}$$

Let m be a fixed quantity and let n diminish without limit: then mn also diminishes without limit. Now, if ϕn have the limit N when n diminishes without limit, $\phi(mn)$ must have the same limit. The only difference is, that (say $m=6$) for any very small value of n, $\phi(6n)$ will not have come so near to its limit as ϕn. For the nature of the limit being, that, by taking n sufficiently small, we may make ϕn within any given fraction (say k, which may be as small as we please) of N, we may, by taking the sixth part of the requisite value for n, make $\phi(6n)$ within the same degree of nearness to N. Hence the limits of the two,

$$\frac{z^{n}-1}{n} \quad \text{and} \quad \frac{z^{mn}-1}{mn}$$

are the same. But if the first be called ψz, the second is $\frac{1}{m}\psi(z^m)$; and we have

$$\text{limit of } \psi z = \text{limit of } \frac{1}{m}\psi z^m$$

$$= \frac{1}{m} \text{ limit of } \psi z^m$$

or

$$z-1-\frac{1}{2}(z-1)^2+\&\text{c.} = \frac{1}{m}\left(z^m-1-\frac{1}{2}(z^m-1)^2+\&\text{c.}\right)$$

a property of the series which will serve to verify succeeding investigations.

We have not yet included in our results the case in which the exponent is incommensurable, such as

$$(1+x)^{\sqrt{2}}$$

but since we look upon $\sqrt{2}$ as the limit to which we approach nearer and nearer in the series (obtained from arithmetical extraction in this case),

$$1 \quad 1{\cdot}4 \quad 1{\cdot}41 \quad 1{\cdot}414 \quad 1{\cdot}4142 \;\&\text{c.}$$

we must regard $(1+x)^{\sqrt{2}}$ as the limit to which we approach by taking the successive expressions

$$1+x \quad (1+x)^{\frac{14}{10}} \quad (1+x)^{\frac{141}{100}} \quad (1+x)^{\frac{1414}{1000}} \quad \&\text{c.}$$

so that, whatever approximation k may be used for $\sqrt{2}$ or the limit for which this symbol stands, then

$$1+kx+k\frac{k-1}{2}x^2+\text{ \&c.}$$

is the corresponding approximation to $(1+x)^{\sqrt{2}}$. When the series is convergent, it is evident that if each term be found, say within its thousandth part, the sum of the whole series is found within its thousandth part. Suppose k and $k+m$ to be two approximations to $\sqrt{2}$, as in page 101, the first too small, the second too great, and suppose we compare the pth terms of the corresponding approximations to $(1+x)^{\sqrt{2}}$, or

$$k\frac{k-1}{2}\cdots\cdot\frac{k-p+2}{p-1}x^{p-1} \text{ and } (k+m)\frac{k+m-1}{2}\cdots\cdot\frac{k+m-p+2}{p-1}x^{p-1}$$

the ratio of which is

$$\frac{k+m}{k}\cdot\frac{k+m-1}{k-1}\cdots\cdots\frac{k+m-p+2}{k-p+2}$$

Since m can be made as small as we please, it is evident that each of these factors can be made as near to unity as we please, and, consequently, their product can be brought as near to unity as we please. That is, for any *given* number of terms, the terms of the two approximations may be made as nearly equal as we please. Now, suppose x less than 1, so that (page 210), both approximations are convergent. Hence so many terms (say q terms) may be taken that the limit of the sum of the remainder shall be as small as we please: and as m may be taken so small that all the q terms of one approximation shall be within, say their millionth parts, of those of the other approximation, it follows that we may reduce the two approximations to the following form:

$a+b+c+\cdots\cdots \quad +z \quad +$ { A remainder less than the millionth of the preceding sum.

$a(1+\alpha)+b(1+\beta)+\cdot\cdot +z(1+\omega)+$ { A remainder less than the millionth of the preceding sum.

where $\alpha\ \beta\ \ldots\ldots\ \omega$ are severally less than one millionth. Let $a+b+c+\ldots\ldots+z$ (which we may call the approximation to the first approximation) be called P; $a\alpha+b\beta+\ldots\ldots+z\omega$ is less than one millionth of P, and the first approximation complete is

$$P + XP \qquad (X \text{ is less than one millionth})$$

and the second

$$P(1+V) + YP(1+V) \qquad (V \text{ and } Y \text{ do. do.})$$

the difference of which is

$$PV + P(Y-X) + PYV$$

which is less than *three* millionths of P; because V, Y — X and YV are severally less than *one* millionth of P. But the limit $(1+x)^{\sqrt{2}}$ must lie between these approximations, and therefore does not differ from P by so much as the approximations differ from each other, that is, by three millionths of P.

It is not possible to shew, by the preceding method, the same result in the case where x is greater than 1, or the series divergent: but it must be remembered that in this case *algebraical* equality only is asserted, not *arithmetical* equality: and all that is said is, that

$$1 + \sqrt{2}x + \sqrt{2}\frac{\sqrt{2}-1}{2}x^2 + \&c.$$

may be substituted without error for $(1+x)^{\sqrt{2}}$; so that the proof of this case falls within the general proof, as deduced from the principle in page 212. What is done in the preceding process shews the approximation to *arithmetical* equality, when the series is convergent.

The same arguments might be applied to any other case, with any other degree of approximation. We now proceed to develope some further consequences of the binomial theorem.

CHAPTER XII.

EXPONENTIAL AND LOGARITHMIC SERIES.

An algebraic symbol may have different names, according to the relation in which it stands to different symbols, or combinations of symbols. Thus in ab, a, with respect to b, is called a *coefficient;* with respect to ab, it is called a *factor*. Similarly, in a^b, b, with respect to a, is called an *exponent;* with respect to a^b, it is called a *logarithm*,* and a in reference to b is called the *base* of the logarithm.

Thus, in 3^4, 4 is the logarithm of 3^4 or 81, to the base 3; in a^x, x is the logarithm of a^x to the base a. This we shall denote by $4 = \log_3 81$ and $x = \log_a a^x$: the letters *log* being an abbreviation of *logarithm*, and the underwritten figure being the base.

Examples. $10^3 = 1000 \qquad 3 = \log_{10} 1000$

If $\quad a^x = y \qquad x = \log_a y$

If $\quad p^q = 1-z \qquad q = \log_p (1-z)$

To construct a system of logarithms to a given base, say 10, we must solve the series of equations

$$10^x = 1 \qquad 10^x = 2 \qquad 10^x = 3 \qquad 10^x = 4 \text{ \&c.}$$

and find the value of x in each. This can, generally speaking, only be done by approximation: that is, the logarithm is generally incommensurable with the unit. By saying, then, that $\log_{10} 2 = \cdot 30103$, we mean that

$$10^{\cdot 30103} = 2 \textit{ very nearly}, \text{ or } \sqrt[100000]{10^{30103}} = 2 \text{ nearly},$$

and that a fraction k can be found such that 10^k shall be as near to 2 as we please, to which fraction $\cdot 30103$ is an approximation.

[In all the following theorems one base is supposed, namely, a.]

Theorem I. *Whatever the base may be, the logarithm of* 1 *is* 0. This is evidently another way of expressing $a^0 = 1$, and may be written $\log_a 1 = 0$.

* From λογων ἀριθμος, the *number of the ratios*, an idea derived from an old method of constructing logarithms, which cannot be here explained.

THEOREM II. *The logarithm of the base itself is* 1. This is contained in $a^1 = a$, and is expressed thus: $\log_a a = 1$

THEOREM III. *The logarithms of* y *and* $\frac{1}{y}$ *are of different signs, but equal numerical value.* For if $y = a^x$ or $x = \log_a y$, we have $\frac{1}{y} = a^{-x}$ or $-x = \log_a \frac{1}{y}$; that is, $\log_a \frac{1}{y} = -\log_a y$.

THEOREM IV. *If* a *be the base, and a number or fraction lie between* a^m *and* a^n, the logarithm of that number or fraction lies between m and n.

For if a^x, the number, lie between a^m and a^n; then x, the logarithm, lies between m and n (see page 89).

BASE 10.			BASE $\frac{1}{2}$.		
Number between	has its	Logarithm between	Number between	has its	Logarithm between
1 and 10		0 and 1	1 and $\frac{1}{2}$		0 and 1
10 and 100		1 and 2	$\frac{1}{2}$ and $\frac{1}{4}$		1 and 2
100 and 1000		2 and 3	$\frac{1}{4}$ and $\frac{1}{8}$		2 and 3
&c.		&c.	&c.		&c.
1 and $\frac{1}{10}$		0 and -1	1 and 2		0 and -1
$\frac{1}{10}$ and $\frac{1}{100}$		-1 and -2	2 and 4		-1 and -2
$\frac{1}{100}$ and $\frac{1}{1000}$		-2 and -3	4 and 8		-2 and -3
&c.		&c.	&c.		&c.

THEOREM V. *The logarithm of a product is equal to the sum of the logarithms of the factors.* Let a be the base, and p, q, and r, the logarithms of P, Q, and R. Then

$$P = a^p \qquad Q = a^q \qquad R = a^r$$

$$PQR = a^{p+q+r} \text{ or } \log(PQR) = p + q + r = \log P + \log Q + \log R$$

THEOREM VI. *The logarithm of a ratio, quotient, or fraction, is the difference of the logarithms of the antecedent and consequent, dividend and divisor, or numerator and denominator.*

For $\frac{P}{Q} = a^{p-q}$ or $\log\frac{P}{Q} = p-q = \log P - \log Q$

THEOREM VII. *The logarithm of* P^m *is found by multiplying the logarithm of* P *by* m; that is, if $P = a^p$, $P^m = a^{mp}$, or $\log P^m = mp = m \log P$.

THEOREM VIII. *A negative number has no arithmetical logarithm: nor is a system of logarithms with a negative base within the limits of algebra as hitherto considered.* This is rather a definition than a theorem,* and it amounts to this: on account of certain anomalies, of which the explanation cannot yet be understood by the student, we are obliged to defer the consideration of the logarithms of negative quantities, and all logarithms of positive quantities, except only those which have arithmetical meaning. For the equation $a^x = b$ has not been proved to have only one solution, though it has only one *arithmetical* solution.

THEOREM IX. *The logarithm of* 0 *is infinite;* by which we mean that as y diminishes without limit, its logarithm (being always negative on one side or other of 1), increases numerically without limit. To diminish $\left(\frac{1}{2}\right)^x$ without limit, x must undergo numerical increase without limit, and be positive: to diminish 2^x without limit, x must also undergo numerical increase without limit, being negative. Hence (page 156) the meaning of the proposition stated; and it must be observed that the symbol $\propto$ has either sign in algebra, which may be explained as follows. If $y = \frac{1}{x}$, then y and x must have the same sign: if x diminish without limit, it approaches a form (0) in which it has no sign, being the limiting boundary between positive and negative quantity. Consequently y, which increases without limit, approaches a similar boundary; for since y has the same sign as $\frac{1}{x}$, if there be any form of x in which either sign may be supposed, the same may be supposed for y. But such a comprehension of this point as can be derived from frequent instances must be reserved until the student has seen more examples of the application of algebra to geometry.

* Being usually considered as a theorem, we have stated it as such.

The following examples will illustrate the preceding theorems.

$$x \times 1 = x \qquad \log x + \log 1 = \log x \qquad (\log 1 = 0)$$

$$x^1 = x \qquad 1 \times \log x \quad = \log x$$

$$\log_a ax = \log_a x + \log_a a = \log_a x + 1$$

$$\log x\sqrt{y} = \log x + \log y^{\frac{1}{2}} = \log x + \frac{1}{2}\log y$$

$$\log \frac{xy^{\frac{1}{3}}}{pq^2} = \log x + \frac{1}{3}\log y - \log p - 2\log q$$

$$\log \left(\frac{xy^3}{pq^{-1}}\right)^{-1} = -1\left\{\log x + 3\log y - \log p - (-1)\log q\right\}$$

$$= -\log x - 3\log y + \log p - \log q$$

We shall now proceed to the series connected with logarithms.

In page 218, we have proved the following theorem for all values of n and x,

$$\left\{1 + 1 + \frac{1-\frac{1}{n}}{2} + \frac{1-\frac{1}{n}}{2}\,\frac{1-\frac{2}{n}}{3} + \&c.\right\}^x = 1 + x + x\frac{x-\frac{1}{n}}{2} + x\frac{x-\frac{1}{n}}{2}\,\frac{x-\frac{2}{n}}{3} + \&c.$$

In which both series, being forms of $\left(1+\frac{1}{n}\right)^q$, will (page 210) be convergent whenever $\frac{1}{n}$ is less than 1, or n greater than 1. Now, let n increase without limit, in which case the limit of

$$1 + 1 + \frac{1-\frac{1}{n}}{2} + \frac{1-\frac{1}{n}}{2}\,\frac{1-\frac{2}{n}}{3} + \&c. \text{ is } 1 + 1 + \frac{1}{2} + \frac{1}{2\,.\,3} + \&c.$$

and that of

$$1 + x + \frac{x-\frac{1}{n}}{2} + \frac{x-\frac{1}{n}}{2}\,\frac{x-\frac{2}{n}}{3} + \&c. \text{ is } 1 + x + \frac{x^2}{2} + \frac{x^3}{2\,.\,3} + \&c.$$

But $1+1+\frac{1}{2}+$ &c. has been calculated approximately in page 183, found to be 2·71828182 ...., and called ε. Therefore, (page 183),

$$\varepsilon^x = 1 + x + \frac{x^2}{2} + \frac{x^3}{2\,.\,3} + \&c.$$

Consequently, *if* ε *be the base, the number which has the logarithm* x *is* $1+x+\frac{x^2}{2}+$ &c. Hence a system of logarithms, having ε for the base, being arrived at in the course of investigation, has been called the *natural* system of logarithms; it is also called the *Naperian** system of logarithms, and the *hyperbolic*† system of logarithms.

And here let it be remembered, that *in algebraical analysis, the letters* log, *by themselves, imply the logarithm to the base* ε, *any exception being always specially mentioned.*

The preceding equation being always true, we have

$$\varepsilon^{kx} = 1 + kx + \frac{k^2x^2}{2} + \frac{k^3x^3}{2.3} + \&c.$$

But ε^{kx} is $(\varepsilon^k)^x$; and if we take k to be the logarithm (base ε) of a, we have $\varepsilon^k = a$, and

$$(\varepsilon^k)^x \text{ or } a^x = 1 + (\log a)x + \frac{(\log a)^2 x^2}{2} + \frac{(\log a)^3 x^3}{2.3} + \&c.$$

From which we find that $\frac{a^x - 1}{x}$ will, if x be diminished without limit, have the limit $\log a$.

But (page 218) we have already found the series for this limit; whence

$$\log a = (a-1) - \frac{1}{2}(a-1)^2 + \frac{1}{3}(a-1)^3 - \&c.$$

or, if $a = 1 + b$, we have

$$\log(1+b) = b - \frac{b^2}{2} + \frac{b^3}{3} - \&c. \;\ldots\ldots\ldots\ldots\; (1)$$

Substitute $-b$ for b, and we have

$$\log(1-b) = -b - \frac{b^2}{2} - \frac{b^3}{3} - \&c. \;\ldots\ldots\ldots\; (2)$$

Subtract (2) from (1) $\left[\log(1+b) - \log(1-b) = \log\left(\frac{1+b}{1-b}\right)\right]$

$$\log\left(\frac{1+b}{1-b}\right) = 2\left\{b + \frac{b^3}{3} + \frac{b^5}{5} + \&c.\right\} \;\cdots\; (3)$$

let $\frac{1+b}{1-b} = \frac{1+x}{x}$ which gives $b = \frac{1}{2x+1}$

* John Napier, commonly called Lord Napier, though not a peer, or otherwise entitled to the appellation of *lord* than in the way in which many landed proprietors are *lords* (of manors), was born in 1550, at his *lordship* of Merchiston, near Edinburgh, and died in 1617. He published the invention of logarithms (his method also leading him to the *natural* system) in 1614.

† So called from a supposed analogy with the curve called the *hyperbola*, but which analogy belongs equally to all systems of logarithms.

$$\log\frac{1+b}{1-b} = \log\frac{1+x}{x} = \log(1+x) - \log x$$

$$\log(x+1) - \log x = 2\left\{\frac{1}{2x+1} + \frac{1}{3}\frac{1}{(2x+1)^3} + \frac{1}{5}\frac{1}{(2x+1)^5} + \&c.\right\} \;..\; (4)$$

The last series gives

$$x = 1 \quad \log 2 = \qquad 2\left\{\frac{1}{3} + \frac{1}{3}\frac{1}{27} + \frac{1}{5}\frac{1}{243} + \&c.\right\}$$

$$x = 2 \quad \log 3 = \log 2 + 2\left\{\frac{1}{5} + \frac{1}{3}\frac{1}{125} + \frac{1}{5}\frac{1}{3125} + \&c.\right\}$$

$$x = 3 \quad \log 4 = \log 3 + 2\left\{\frac{1}{7} + \frac{1}{3}\frac{1}{343} + \frac{1}{5}\frac{1}{16807} + \&c.\right\}$$

$$x = 4 \quad \log 5 = \log 4 + 2\left\{\frac{1}{9} + \frac{1}{3}\frac{1}{729} + \frac{1}{5}\frac{1}{59049} + \&c.\right\}$$

Thus the logarithms of whole numbers may be successively calculared with tolerable readiness, as the first example (which contains the least convergent series), here given at length to ten places (that is to eleven for the sake of accuracy), will shew.

$$\frac{1}{3} = \cdot 33333333333$$

$$\frac{1}{3}\frac{1}{3^3} = \cdot 01234567901$$

$$\frac{1}{5}\frac{1}{3^5} = \cdot 00082304527$$

$$\frac{1}{7}\frac{1}{3^7} = \cdot 00006532105$$

$$\frac{1}{9}\frac{1}{3^9} = \cdot 00000564503$$

$$\frac{1}{11}\frac{1}{3^{11}} = \cdot 00000051318$$

$$\frac{1}{13}\frac{1}{3^{13}} = \cdot 00000004825$$

$$\frac{1}{15}\frac{1}{3^{15}} = \cdot 00000000465$$

$$\frac{1}{17}\frac{1}{3^{17}} = \cdot 00000000046$$

$$\frac{1}{19}\frac{1}{3^{19}} = \cdot 00000000005$$

$$\cdot 34657359028$$

$$2$$

$$\log 2 = \cdot 69314718056 \text{ very nearly.}$$

By this means the logarithm of 3 may be found from that of 2, that of 4 from that of 3, and so up to any given whole number. It would be desirable,* as an exercise of arithmetic, that the student should calculate them up to 10 inclusive; the results of which (to eight places) would be as follows:

log 1 = 0·00000000	log 6 = 1·79175947
log 2 = 0·69314718	log 7 = 1·94591015
log 3 = 1·09861229	log 8 = 2·07944154
log 4 = 1·38629436	log 9 = 2·19722458
log 5 = 1·60943791	log 10 = 2·30258509

But the series need be employed only for prime numbers, and the first tables of logarithms were thus constructed, as follows. Suppose the logarithm of 59 to be required, or of $58+1$. Now 58 is 2×29, both factors being prime numbers; if, then, we have the logarithms of 2 and 29, we have that of 58 from the equation

$$\log 58 = \log 2 + \log 29$$

and that of 59 from

$$\log 59 = \log 58 + 2\left\{\frac{1}{117} + \frac{1}{3}\frac{1}{(117)^3} + \&c.\right\}$$

Beginning, then, with log 2, we have the following:

log 2 = a given series	log 6 = log 3 + log 2
log 3 = log 2 + a given series	log 7 = log 6 + a given series
log 4 = 2 log 2	log 8 = 3 log 2
log 5 = log 4 + a given series	log 9 = 2 log 3

log 10 = log 2 + log 5; and so on.

* Halley (*Phil. Trans.* 1695) thus expresses himself, after having described the preceding method. "If the curiosity of any gentleman that has leisure, would prompt him to undertake to do the logarithms of all prime numbers under a hundred thousand to twenty-five or thirty places of figures, I dare assure him that the facility of this method will invite him thereto; nor can any thing more easy be desired." Without insisting upon any thing that would take up so much of a gentleman's leisure as the preceding, I should strongly recommend the student always to work one example of every moderately convergent series which occurs.

In page 219, we proved independently that

$$z-1-\frac{1}{2}(z-1)^2+\ \&c. = \frac{1}{m}\left(z^m-1-\frac{1}{2}(z^m-1)^2+\ \&c.\right)$$

which is, as we now see, the same thing as

$$\log z = \frac{1}{m}\log z^m$$

and in it we also find further elucidation of the equation,

$$\text{limit of } \frac{z^m-1}{m} = z-1-\frac{1}{2}(z-1)^2+\ \&c. = \log z.$$

The diminution of m without limit, or the supposing of m to be a smaller and smaller fraction, implies the extraction of higher and higher roots of z. By extracting a sufficiently high root of z, we can bring z^m as near to 1 as we please, or make z^m-1 as small as we please; that is (page 187) z^m-1 may be made as nearly equal to the sum of the whole series as we please. It was from this principle, by continual extraction of the square root of z, that logarithms were once calculated, by means of the formula

$$\log z = \left(z^{\frac{1}{140737488355328}}-1\right)\times 140737488355328$$

very nearly,* the number named being 2^{47}, equivalent to 47 extractions of the square root.

The logarithm of any whole number being found, as in the last page, that of a fraction can then be found by the subtraction of the logarithm of the denominator from that of the numerator.

We also notice the following result: when x is a large number,

$$\log(x+1) = \log x + \frac{2}{2x+1} \text{ nearly (page 227).}$$

The rest of this subject will be reserved for the next chapter, on the practical use of logarithms in shortening arithmetical computations. We now proceed with some uses of the preceding series.

Lemma. If $f(x)$ be such a function of x, that $f(x+y)$ can be expanded in a series of the form

$$A_0+A_1y+A_2y^2+\ \&c.$$

* Halley, in the memoir already cited. Each square root was extracted to 14 places.

where A_0, A_1, &c. are functions of x *only*, then

$$f(a+b\sqrt{-1})+f(a-b\sqrt{-1})$$

treated by common rules, will always represent a quantity, either positive or negative, that is, all purely symbolical or *impossible** quantities disappear; while, on the other hand,

$$f(a+b\sqrt{-1})-f(a-b\sqrt{-1})$$

will be of the form (possible quantity) $\times\sqrt{-1}$

Write the value of $f(x+y)$ and afterwards that of $f(x-y)$, made by changing the sign of y;

$$f(x+y) = A_0 + A_1 y + A_2 y^2 + A_3 y^3 + \text{ \&c.}$$
$$f(x-y) = A_0 - A_1 y + A_2 y^2 - A_3 y^3 + \text{ \&c.}$$

from which we find,

$$f(x+y)+f(x-y) = 2A_0 \quad + 2A_2 y^2 + 2A_4 y^4 + \text{ \&c.}$$
$$f(x+y)-f(x-y) = 2A_1 y + 2A_3 y^3 + 2A_5 y^5 + \text{ \&c.}$$

for x write a, whence A_0, A_1, &c. become functions of a only; for y write $b\sqrt{-1}$, that is, suppose

$$y = b\sqrt{-1} \qquad\qquad y^5 = b^5\sqrt{-1}$$
$$y^2 = b^2\times -1 = -b^2 \qquad\qquad y^6 = -b^6$$
$$y^3 = -b^2\times b\sqrt{-1} = -b^3\sqrt{-1} \qquad y^7 = -b^7\sqrt{-1}$$
$$y^4 = -b^3\sqrt{-1}\times b\sqrt{-1} \qquad\qquad y^8 = b^8 \text{ \&c.}$$
$$= -b^4\times -1 = b^4$$

whence

$$f(a+b\sqrt{-1})+f(a-b\sqrt{-1}) = 2A_0 - 2A_2 b^2 + 2A_4 b^4 - \text{ \&c.}$$

which is a possible quantity: and

$$f(a+b\sqrt{-1})-f(a-b\sqrt{-1}) = 2A_1 b\sqrt{-1} - 2A_3 b^3\sqrt{-1} + \text{ \&c.}$$
$$= \left\{2A_1 b - 2A_3 b^3 + \text{ \&c.}\right\}\sqrt{-1}$$

which is (a possible quantity) $\times\sqrt{-1}$.

* We shall now begin to make use of this common phrase: to the student it must mean "impossible till further explained." See page 110.

EXAMPLES. $$\frac{1}{a+b\sqrt{-1}}+\frac{1}{a-b\sqrt{-1}}=\frac{2a}{a^2+b^2}$$

$$\frac{1}{a+b\sqrt{-1}}-\frac{1}{a-b\sqrt{-1}}=-\frac{2b}{a^2+b^2}\sqrt{-1}$$

$$(a+b\sqrt{-1})^n+(a-b\sqrt{-1})^n=2a^n-2n\frac{n-1}{2}a^{n-2}b^2+\&c.$$

$$(a+b\sqrt{-1})^n-(a-b\sqrt{-1})^n=$$
$$\left(2na^{n-1}b-2n\frac{n-1}{2}\,\frac{n-2}{3}\,a^{n-3}b^3+\&c.\right)\sqrt{-1}$$

As the preceding theorem is true for all values of a, it is true when $a=0$; that is, for

$$f(b\sqrt{-1})+f(-b\sqrt{-1})\text{ and }f(b\sqrt{-1})-f(-b\sqrt{-1})$$

EXAMPLES. Let n be a positive whole number.

$$(b\sqrt{-1})^n+(-b\sqrt{-1})^n=\begin{cases}+2b^n \text{ when } n \text{ is \textit{evenly even}.*}\\ \;\;0 \quad\text{ when } n \text{ is odd.}\\ -2b^n \text{ when } n \text{ is \textit{oddly even}.†}\end{cases}$$

$$(b\sqrt{-1})^n-(-b\sqrt{-1})^n=\begin{cases}0 \text{ when } n \text{ is even.}\\ 2b^n\sqrt{-1} \text{ when } n \text{ is } 1, 5, 9, \&c.\\ -2b^n\sqrt{-1} \text{ when } n \text{ is } 3, 7, 11, \&c.\end{cases}$$

If we apply the same process to $\varepsilon^{x\sqrt{-1}}$ and $\varepsilon^{-x\sqrt{-1}}$, we find

$$\varepsilon^{x\sqrt{-1}}=1+x\sqrt{-1}-\frac{x^2}{2}-\frac{x^3}{2.3}\sqrt{-1}+\frac{x^4}{2.3.4}+\&c.$$

$$\varepsilon^{-x\sqrt{-1}}=1-x\sqrt{-1}-\frac{x^2}{2}+\frac{x^3}{2.3}\sqrt{-1}+\frac{x^4}{2.3.4}-\&c.$$

$$\frac{\varepsilon^{x\sqrt{-1}}+\varepsilon^{-x\sqrt{-1}}}{2}=1-\frac{x^2}{2}+\frac{x^4}{2.3.4}-\frac{x^6}{2.3.4.5.6}+\&c. \quad \ldots. \text{ (A)}$$

$$\frac{\varepsilon^{x\sqrt{-1}}-\varepsilon^{-x\sqrt{-1}}}{2\sqrt{-1}}=x-\frac{x^3}{2.3}+\frac{x^5}{2.3.4.5}-\&c. \quad \ldots. \text{ (B)}$$

We have left the use of the symbol $\sqrt{-1}$, in this work, to be justified by experience only (see page 111); we have now an oppor-

* Divisible by 4; 4, 8, 12, &c. are evenly even.
† Divisible by 2, but not by 4; as 2, 6, 10, &c.

tunity of examining the result of a long series of deduction, with a view of ascertaining how far we shall produce consistent results. The algebraical sign of equality is placed between the two sides of the preceding equations; the question is, Will any relations we may discover to exist between the two *first* sides also exist between the two *second* sides?

Let $$= \frac{\varepsilon^{x\sqrt{-1}} + \varepsilon^{-x\sqrt{-1}}}{2} \text{ be called } \varphi x$$

and $$\frac{\varepsilon^{x\sqrt{-1}} - \varepsilon^{-x\sqrt{-1}}}{2\sqrt{-1}} \text{ be called } \psi x$$

we have then

$$(\varphi x)^2 = \frac{\varepsilon^{2x\sqrt{-1}} + 2\varepsilon^{x\sqrt{-1}}\varepsilon^{-x\sqrt{-1}} + \varepsilon^{-2x\sqrt{-1}}}{4}$$

$$(\psi x)^2 = \frac{\varepsilon^{2x\sqrt{-1}} - 2\varepsilon^{x\sqrt{-1}}\varepsilon^{-x\sqrt{-1}} + \varepsilon^{-2x\sqrt{-1}}}{-4}$$

$$(\varphi x)^2 + (\psi x)^2 = \frac{4\varepsilon^{x\sqrt{-1}-x\sqrt{-1}}}{4} = \varepsilon^0 = 1$$

$$(\varphi x)^2 - (\psi x)^2 = \frac{\varepsilon^{2x\sqrt{-1}} + \varepsilon^{-2x\sqrt{-1}}}{2} = \varphi(2x)$$

$$\varphi x \times \psi x = \frac{\varepsilon^{2x\sqrt{-1}} - \varepsilon^{-2x\sqrt{-1}}}{4\sqrt{-1}} = \frac{1}{2}\psi(2x)$$

of which three relations, namely,

$$(\varphi x)^2 + (\psi x)^2 = 1 \qquad (\varphi x)^2 - (\psi x)^2 = \varphi(2x)$$
$$2\varphi x \times \psi x = \psi(2x)$$

it is asked, are they true of the second sides of (A) and (B)?

The multiplication of a series by itself will be found to amount to using the following rule: square each term, and multiply all that follow by twice that term. Thus, the square of the series in (A) is

$$1 - x^2 + \frac{x^4}{3.4} - \frac{x^6}{3.4.5.6} + \&c.$$
$$+ \frac{x^4}{2^2} - \frac{x^6}{2.3.4} + \&c.$$
$$+ \&c.$$

that of the series in (B) is

$$x^2 - \frac{x^4}{3} + \frac{x^6}{3\,.\,4\,.\,5} - \&c.$$
$$+ \frac{x^6}{2^2\,.\,3^2} - \&c.$$
$$- \&c.$$

The first square increased by the second is

$$1 + \left\{\frac{1}{3\,.\,4} + \frac{1}{2^2} - \frac{1}{3}\right\}x^4 - \left\{\frac{1}{3\,.\,4\,.\,5\,.\,6} + \frac{1}{2\,.\,3\,.\,4} - \frac{1}{3\,.\,4\,.\,5} - \frac{1}{2^2\,3^2}\right\}x^6 + \&c.$$
$$= 1 + \{\ 0\ \} \qquad - \{\ 0\ \} \qquad + \&c.$$

The first diminished by the second is

$$1 - 2x^2 + \left\{\frac{1}{3\,.\,4} + \frac{1}{2^2} + \frac{1}{3}\right\}x^4 - \left\{\frac{1}{3\,.\,4\,.\,5\,.\,6} + \frac{1}{2\,.\,3\,.\,4} + \frac{1}{3\,.\,4\,.\,5} + \frac{1}{2^2 3^2}\right\}x^6 + \&c.$$
$$= 1 - \frac{(2x)^2}{2} + \frac{(2x)^4}{2\,.\,3\,.\,4} - \frac{(2x)^6}{2\,.\,3\,.\,4\,.\,5\,.\,6} + \&c.$$

The third relation may be similarly verified by multiplication.

The following results may also be proved in the same way both of the first and second sides of equations (A) and (B)

$$\phi(x+y) = \phi x\,\phi y - \psi x\,\psi y \qquad \psi(x+y) = \psi x\,\phi y + \phi x\,\psi y$$
$$\phi(x-y) = \phi x\,\phi y + \psi x\,\psi y \qquad \psi(x-y) = \psi x\,\phi y - \phi x\,\psi y$$

Let $\varepsilon^{x\sqrt{-1}}$ be called p, and let $\frac{\psi x}{\phi x}$ be called χx, then (equations A and B)

$$\frac{1}{\sqrt{-1}}\,\frac{p - \frac{1}{p}}{p + \frac{1}{p}} = \frac{\psi x}{\phi x} = \chi x \quad \text{or} \quad \frac{p^2 - 1}{p^2 + 1} = \sqrt{-1}\,\chi x$$

whence $\qquad p^2 = \dfrac{1 + \sqrt{-1}\,\chi x}{1 - \sqrt{-1}\,\chi x} \qquad$ and, page 226,

$$\log p^2 = 2\left\{\sqrt{-1}\,\chi x + \frac{1}{3}\left(\sqrt{-1}\,\chi x\right)^3 + \&c.\right\}$$
$$= 2\sqrt{-1}\left\{\chi x - \frac{1}{3}\left(\chi x\right)^3 + \frac{1}{5}\left(\chi x\right)^5 - \&c.\right\}$$

But $\quad p^2 = \varepsilon^{2x\sqrt{-1}}$ or $\log p^2 = 2x\sqrt{-1}$, therefore

$$x = \chi x - \frac{1}{3}\left(\chi x\right)^3 + \frac{1}{5}\left(\chi x\right)^5 - \&c.$$

The series A and B are always convergent, as may be proved by the test in page 185; but the convergency may be made to begin at any term, however distant, by making x sufficiently great. Thus, if x were 1000, the first series would not begin to converge before the term

$$\frac{x^{264}}{2.3.4.5\ldots\ldots\ldots 263.264}$$

but, notwithstanding the magnitude of the first terms of the series (which it must be remembered, are alternately added and subtracted) the actual value of ϕx and ψx can never exceed 1; for if either were arithmetically greater than 1 (positive or negative), the equation $(\phi x)^2 + (\psi x)^2 = 1$ could not be true.

In trigonometry, the properties of the preceding series are connected with geometry in the following way. Let a circle be drawn;

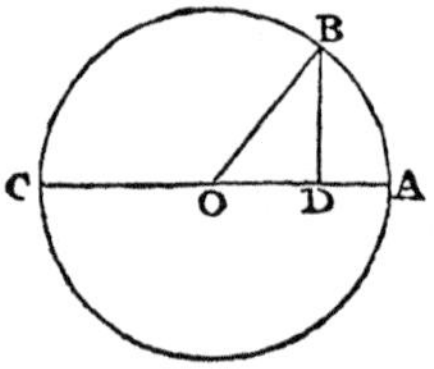

and from A let the point B set out until it has described an arc equal in length to x times the radius OA, going round the circle again if necessary. Draw BD perpendicular to CA; then it will be proved that BD is the fraction ψx of OA, and OD the fraction ϕx of OA.

EXERCISE 1. By means of the equation

$\dfrac{1+x}{1+\frac{1}{x}} = x$, prove the equation

$$\log x = x - \frac{1}{x} - \frac{1}{2}\left(x^2 - \frac{1}{x^2}\right) + \frac{1}{3}\left(x^3 - \frac{1}{x^3}\right) - \&c.$$

2. Prove the following,

$$\varepsilon^{x\sqrt{-1}} = \varphi x + \sqrt{-1}\,\psi x$$

$$\varepsilon^{-x\sqrt{-1}} = \varphi x - \sqrt{-1}\,\psi x$$

$$(\varphi x + \sqrt{-1}\,\psi x)^m = \varphi(mx) + \sqrt{-1}\,\psi(mx)$$

CHAPTER XIII.

ON THE USE OF LOGARITHMS IN FACILITATING COMPUTATIONS.

THE natural system of logarithms, already explained, has this defect as an instrument of calculation, that there is no method of finding the logarithm of a fraction more simple than subtracting the logarithm of the denominator from that of the numerator. For instance, the logarithm of ·3 can not be more shortly found than by taking

$$\log 3 - \log 10 \quad \text{or} \quad 1{\cdot}09861229 - 2{\cdot}30258509 = -1{\cdot}20397280$$

We proceed to find another system in which there shall be some more obvious connexion between the logarithm of 3 and those of ·3, ·03, &c.

The *fundamental* definition of $\log_a x$ gives

$$x = a^{\log_a x} \qquad\qquad a = x^{\frac{1}{\log_a x}}$$

We have then $\quad x = a^{\log_a x} = b^{\log_b x}$

But $\quad b = a^{\log_a b} \qquad \therefore x = a^{\log_a b \log_b x}$

Similarly $\quad x = a^{\log_a c \log_c b \log_b x}$

$$= a^{\log_a e \log_e c \log_c b \log_b x}$$

which last result may be also thus verified:

$$a^{\log_a e \log_e c \log_c b \log_b x} = \left\{a^{\log_a e}\right\}^{\log_e c \log_c b \log_b x}$$

$$= e^{\log_e c \log_c b \log_b x} = c^{\log_c b \log_b x} = b^{\log_b x} = x$$

Now, there is but one arithmetical exponent which applied to a will give x; for, if possible, let there be two, p and q, and let $a^p = x$, $a^q = x$, whence $a^p = a^q$, and $a^{p-q} = 1$; therefore $p - q = 0$, or $p = q$, that is, p and q do not differ, as was supposed. Hence, since

$$x = a^{\log_a x} = a^{\log_a b \log_b x}, \text{ we have}$$

$$\log_a x = \log_a b \log_b x = \log_a c \log_c b \log_b x \text{ \&c.}$$

these results may be remembered by means of the identical equations

$$\frac{x}{a} = \frac{x}{b}\frac{b}{a} \qquad \frac{x}{a} = \frac{x}{b}\frac{b}{c}\frac{c}{a}$$

giving the following theorem : *the series of equations*

$$\frac{a}{b} = \frac{a}{c}\frac{c}{b} = \frac{a}{e}\frac{e}{c}\frac{c}{b} = \frac{a}{f}\frac{f}{e}\frac{e}{c}\frac{c}{b} \text{ \&c.}$$

remains true, if for each fraction be substituted the logarithm of its numerator to the denominator as a base.

As an example, by means of

$\frac{a}{b}\frac{b}{a} = \frac{a}{a}$ we remember that $\log_b a \log_a b = \log_a a = 1$

or
$$\log_a b = \frac{1}{\log_b a}$$

We also have
$$\log_b x = \frac{\log_a x}{\log_a b}$$

or; *to convert a given system of logarithms having the base* a, *into another which shall have the base* b, *divide every logarithm given by the given logarithm of* b.

Seeing that in practice it is convenient to reduce all fractions to decimal fractions, the base chosen should be one in which the logarithms of 10, 100, &c. are whole numbers, that is, it should be 10. For in that case we have

$$\log 10 = 1, \quad \log 100 = 2, \quad \log 1000 = 3, \text{ \&c.}$$

And if we call p the logarithm of any number, say 25, we have

$$\begin{aligned}
\log 2{\cdot}5 &= \log 25 - \log 10 = p-1 \\
\log {\cdot}25 &= \log 25 - \log 100 = p-2 \\
\log {\cdot}025 &= \log 25 - \log 1000 = p-3 \text{ \&c.} \\
\log 250 &= \log 25 + \log 10 = p+1 \\
\log 2500 &= \log 25 + \log 100 = p+2 \text{ \&c.}
\end{aligned}$$

So that, *when the base is ten*, any alteration of the place of the decimal point in the number requires only the addition or subtraction of a whole number from the logarithm.

The system of logarithms to the base 10 is deduced from the natural system by the following equation,

$$\log_{10} x = \frac{\log_\epsilon x}{\log_\epsilon 10} = \frac{\log_\epsilon x}{2{\cdot}30258509} = \log_\epsilon x \times {\cdot}4342944819$$

This system of logarithms is called the *common*, *tabular*, *decimal*, or *Brigg's* system, and ·43429 .... is called its *modulus*, and generally $1 \div \log_\epsilon a$, or $\log_a \epsilon$ is called the modulus of the system whose base is a.

All the logarithms in the remainder of this chapter are common *logarithms*.

The following are examples of the arrangements of some tables of logarithms, for the purpose of explaining how to find the logarithms of given numbers, or *vice versâ*.

1. *Lalande.**

Nomb.	Logarit.	D	Nomb.	Logarit.	D
1080	3.03342	41	1110	3.04532	39
1081	3.03383	40	1111	3.04571	39
1082	3.03423	40	1112	3.04610	40
1083	3.03463	40	1113	3.04650	39
1084	3.03503		1114	3.04689	
&c.	&c.	&c.	&c.	&c.	&c.

2. *Sherwin, Hutton, Babbage.*†

Num.	0	1	2	3	4	5	6	7	8	9	Diff.
5150	7118072	8157	8241	8325	8410	8494	8578	8663	8747	8831	84
1	8915	9000	9084	9168	9253	9337	9421	9506	9590	9674	1 8 2 17
2	9759	9843	9927	0011	0096	0180	0264	0349	0433	0517	3 25 4 34
3	7120601	0686	0770	0854	0939	1023	1107	1191	1276	1360	5 42 6 50
4	1444	1528	1613	1697	1781	1865	1950	2034	2118	2202	7 59 8 67
&c.	&c.	&c.	&c.	&c.	&c.	&c.	&c.	&c.	&c.	&c.	9 76

* *Tables de Logarithmes*, &c. par Jérome de LA LANDE, édition stéréotype, par FIRMIN DIDOT: Paris, chez Firmin Didot, &c. Rue Jacob, No. 24, 1805 (tirage de 1831). This work is sufficient for most purposes, but those who order it should remember to insist on having one of the later *tirages*.

† All these works are well known in this country. The first (an old work) is frequently to be found with second-hand booksellers. The last two can be obtained in the usual way from any bookseller.

A logarithm usually consists of a whole number, followed by a decimal fraction, both or either of which may be negative: but in *most* tables nothing is put down but *positive decimal fractions.* We proceed to shew how this arrangement includes all cases.

1. Take a negative logarithm, such as $-3{\cdot}16804$, which is $-3-{\cdot}16804$, or $-4+(1-{\cdot}16804)$, or $-4+{\cdot}83196$. This is usually written $\bar{4}{\cdot}83196$, in which the negative sign *over* the four means that that figure only is negative. [According to analogy, $\bar{1}3$ would mean $-10+3$, or -7; $1\bar{3}6$ would mean $106-30$, or 76.] Thus, every negative logarithm may be so converted as to have a positive decimal part.

2. When the number corresponding to the decimal part is known, that corresponding to the whole can be immediately found by the following table,

$\log \frac{1}{1000}$ or $\log 10^{-3} = -3$	$\log 10$ or $\log 10^1 = 1$
$\log \frac{1}{100}$ or $\log 10^{-2} = -2$	$\log 100$ or $\log 10^2 = 2$
$\log \frac{1}{10}$ or $\log 10^{-1} = -1$	$\log 1000$ or $\log 10^3 = 3$
	&c.

Thus, the number to ${\cdot}30103$ is 2 very nearly; that is, ${\cdot}30103 = \log 2$: therefore, $1{\cdot}30103 = \log 10 + \log 2 = \log 20$, $\bar{1}{\cdot}30103 = \log\frac{1}{10} + \log 2 = \log\frac{2}{10} = \log{\cdot}2$, and so on.

If D simply stand for a decimal fraction less than unity, we have (page 223) the following table:

The log. of a No. lying between	Must lie between	And must therefore be of the form	Instances of such numbers
$\frac{1}{1000}$ and $\frac{1}{100}$	-3 and -2	$-3+D$	${\cdot}0013$, ${\cdot}0098$
$\frac{1}{100}$ and $\frac{1}{10}$	-2 and -1	$-2+D$	${\cdot}014$, ${\cdot}0738$
$\frac{1}{10}$ and 1	-1 and 0	$-1+D$	${\cdot}103$, ${\cdot}4296$
1 and 10	0 and 1	$0+D$	2·56, 7·99
10 and 100	1 and 2	$1+D$	11·03, 45·96
100 and 1000	2 and 3	$2+D$	159, 159·108
&c.	&c.	&c.	&c.

Definition. The whole part of a common logarithm, whether positive or negative, is called the *characteristic* of the logarithm. The decimal part is called the *mantissa.** We now give some theorems which obviously follow from what precedes.

1. No alteration in the place of the decimal point (in which is included the annexation of ciphers to a whole number) alters the *mantissa* of the logarithm, but only the *characteristic.*

2. When the decimal point of the number *is* preceded by significant figures, the characteristic of the logarithm is *positive,* and *a unit less* than the *number* of these *figures.* Thus, the logarithm of 12345·67 is 4 + *mantissa;* that of 6·9 is 0 + *mantissa.*

3. When the decimal point of the number *is not* preceded by significant figures, the characteristic of its logarithm is *negative,* and *a unit more* than the number of ciphers which precedes the first significant figure. Thus the logarithm of ·00083 is —4 + *mantissa,* that of ·83 is —1 + *mantissa.*

Lalande's table first mentioned gives the characteristic, on the supposition that the number mentioned is a whole number; thus, the logarithm of 1081 (as given in the preceding specimen) is 3·03383. But, to take the logarithm of 1·081 from this table, the mantissa ·03383 is all that must be taken, and the characteristic 0 applied. Thus, the logarithm of 1·081 is 0·03383, and, from the rule laid down, we have the following table:

log 1081000	=	6·03383	log 1·081	=	0·03383
log 108100	=	5·03383	log ·1081	=	$\bar{1}$·03383
log 10810	=	4·03383	log ·01081	=	$\bar{2}$·03383
log 1081	=	3·03383	log ·001081	=	$\bar{3}$·03383
log 108·1	=	2·03383	log ·0001081	=	$\bar{4}$·03383
log 10·81	=	1·03383	log ·00001081	=	$\bar{5}$·03383

The second table gives the logarithms of numbers of five places of figures. From the first table we might have found the logarithms of 5153 and 5154, or (the difference being only in the characteristic) of 51530 and 51540; from the second specimen we can find the

* This word is now seldom used, though there is not another single word which means the same thing.

logarithms of 51531, 51532, ... 51539, intermediate to the two numbers just cited.

Now we must observe, that when a figure is changed in a number, the first figure which changes in the logarithm will be nearer to or further from the left hand, according to the figure changed in the number. This amounts to saying that the smaller (in proportion to the whole) the change of the number, the smaller the change in the logarithm, and is shewn by the following theorem. Since (pages 227, 237) the common logarithm of $1+x$ is ($M = \cdot 43429 \ldots.$)

$$\log(1+x) = \log x + 2M\left(\frac{1}{2x+1} + \frac{1}{3}\frac{1}{(2x+1)^3} + \&c.\right)$$

the greater x is, the less the addition to log x by which log $(x+1)$ is formed. The following instances will illustrate this, in which the figure undergoing change is marked with an accent. The columns succeeding shew, 1st. By how much of itself the number is changed; 2d. By how much of *a unit* the logarithm is changed (nearly),

	Proportion to the whole of the change in the Number.	Absolute change in the Logarithm.
log 1′ = 0·0000000 log 2′ = 0·3010300	1	$\frac{1}{3}$
log 10′ = 1·0000000 log 11′ = 1·0413927	$\frac{1}{10}$	$\frac{1}{25}$
log 100′ = 2·0000000 log 101′ = 2·0043214	$\frac{1}{100}$	$\frac{1}{250}$
log 1000′ = 3·0000000 log 1001′ = 3·0004341	$\frac{1}{1000}$	$\frac{1}{2500}$
log 10000′ = 4·0000000 log 10001′ = 4·0000434	$\frac{1}{10000}$	$\frac{1}{25000}$
log 100000′ = 6·0000000 log 100001′ = 6·0000043	$\frac{1}{100000}$	$\frac{1}{250000}$

Hence, roughly speaking, the absolute change in the mantissa of the logarithm is something less than one half of the relative change in the number. Let the student try to ascertain this from the series given above. Thus, if a number increase by its thousandth part, the increase in the logarithm is less than the absolute fraction $\frac{1}{2000}$, and so on. Hence we can ascertain with sufficient precision to how many places of logarithmic figures it will be necessary to carry any table. Let us suppose, for instance, our table is to give every number of five places, from 10000 to 99999. At the end of the table the relative increase of the number is about $\frac{1}{100000}$, the absolute increase of the logarithm is therefore about $\frac{1}{250000}$, or ·000004. Consequently six places of logarithmic figures are *absolutely* necessary. With less than six places distinction would be lost; for instance,

$$\log 99846 = 5{\cdot}9993307$$
$$\log 99847 = 5{\cdot}9993350$$

which only differ in the *sixth* place. In the first part of the table, where the relative increase is little more than $\frac{1}{10000}$, the absolute increase of the logarithm is nearly ·00004, or five places only would be sufficient. But the tables must, for reasons of practical convenience, be of the same number of figures throughout, and, therefore, must be at least of six places of figures.

In the second specimen, *five* places of figures in the number are accompanied by *seven* places of figures in the logarithms. But, as the three first places of the logarithm continue the same for some time, even in the most changeable part of the table, they are placed by themselves in the first column, at the point where a change takes place: which saves much room, but is subject to this inconvenience, that as a change in the third figure of the logarithm will seldom take place exactly at the beginning of a line, the new *third* figure cannot be shewn till after it has really made its first appearance. The following instances, taken out of the second specimen, will shew both the arrangement of the tables and this new difficulty, better than ny verbal explanation.

Number.	Mant. of the log.	Number.	Mant. of the log.
51520	·711 9759	51526	·712 0264*
51521	·711 9843	51527	·712 0349*
51522	·711 9927	51528	·712 0433*
51523	·712 0011*	51529	·712 0517*
51524	·712 0096*	51530	·712 0601
51525	·712 0180*	51531	·712 0686

In this list it will be observed that each logarithm differs from the preceding either by ·0000083, 0000084, or ·0000085. In fact, the whole difference between the logarithms of 51520 and 51520 + 10 is ·0000842, giving for each increase of a unit an average increase of ·0000084. We have then, at and near 51520, the following equations:

$$\log(51520 + 1) = \log 51520 + \cdot 0000084$$
$$\log(51520 + 2) = \log 51520 + \cdot 0000084 \times 2$$

or, if h be not greater than 10,

$$\log(51520 + h) = \log 51520 + \cdot 0000084 \times h \ldots\ldots (A)$$

or, for a small part of the tables, the logarithms of numbers increasing by a unit increase in arithmetical progression very nearly. Now (M being ·43429 .... pages 226 and 237),

$$\text{Com.}\log\left(1 + \frac{h}{x}\right) = \text{M}\left(\frac{h}{x} - \frac{1}{2}\left(\frac{h}{x}\right)^2 + \frac{1}{3}\left(\frac{h}{x}\right)^3 - \&\text{c.}\right)$$

$$= \text{M}\frac{h}{x} \text{ very nearly, when } \frac{h}{x} \text{ is small (page 187)}$$

or $$\log(x + h) = \log x + \text{M}\frac{h}{x}, \text{ very nearly,}$$

an equation of the same form as (A); whence it follows that (A), which is nearly true when h is 10, is even still more nearly true when h is a fraction of a unit. Hence we have

In this case $x = 51520$ and $\frac{\text{M}}{x}$ or $\frac{\cdot 4342945}{51520} = \cdot 0000084$

* In all these, the first three figures of the mantissa must be looked for below. There are various devices in different tables for reminding the reader of this, which we need not explain, as they are evident on inspection.

$$\log 51250\tfrac{1}{2} = \log 51520 + \cdot 0000084 \times \tfrac{1}{2}$$
$$\log 51250 \cdot 36 = \log 51520 + \cdot 0000084 \times \cdot 36$$

The column marked *Diff.* (for *difference*) is meant to expedite the multiplication which the last equation shews will become necessary when the logarithm of six or seven places is sought. It consists of the *tenths* of 84 to the nearest whole number: thus,

one tenth of 84 is	8·4,	nearest whole number	8
two tenths	16·8,	....	17
three	25·2,	....	25

and so on. The *hundredths* of 84 may be got by striking off one figure from the corresponding *tenths* (adding 1 where the figure struck out is 5 or upwards), thus,

one hundredth of 84 is about	·8,	nearest whole number	1
two hundredths	1·7,	....	2
three	2·5,	....	3

and so on. Hence by inspection of the column of differences we can immediately determine the tenths or hundredths of the difference in question. And now let us determine the logarithm of 51·53946 from the second specimen in page 237.

The *mantissa* is the same as that of the logarithm of 51539·46 and

$$\log(51539 + \cdot 46) = \log 51539 + \cdot 0000084 \times \cdot 46$$
$$= \log 51539 + \cdot 0000084 \left(\tfrac{4}{10} + \tfrac{6}{100}\right)$$

but

$$\cdot 0000084 \left(\tfrac{4}{10} + \tfrac{6}{100}\right) = \cdot 0000001 \left(\tfrac{4}{10} \times 84 + \tfrac{6}{100} \times 84\right)$$
$$\text{(from the table)} \quad = \cdot 0000001\,(34 + 5)$$

But multiplying a whole number by ·1, ·01, ·001, &c. is the same as removing its unit's place to the first, second, third, &c. place of decimals: from which we have

Mantissa of log 51539 from the table	=	·7121360
Addition on account of the 4 which follows the 9	=	34
Ditto on account of the 6 which follows the 4	=	5
Sum		·7121399

This sum is the mantissa of the logarithm of 51539·46, which has the same mantissa as that of 51·53946; therefore, taking the proper characteristic for the last, we have

$$\log 51{\cdot}53946 = 1{\cdot}7121399$$

The following are other instances derived from the same rule, and falling within the limits of the specimen.

log ·5152748?

log ·51527	$\bar{1}$·7120349
4	34
8	7
log ·5152748	$\bar{1}$·7120390

log 5 150008?

log 5·1500	0·7118072
0	00
8	7
log 5·150008	0·7118079

log 5152768000

log 5152700000	9·7120349
6	50
8	7
log 5152768000	9·7120406

log ·00005154899

log ·000051548	$\bar{5}$·7122118
9	76
9	8
log ·00005154899	$\bar{5}$·7122202

The inverse of this question is done as follows: Suppose it required to find from the table the number corresponding to the logarithm $\bar{1}$·7118366. Rejecting the characteristic we look in the table for the mantissa which is nearest to 7118366 (but below it). This we find to be 7118325, which is the logarithm of 51503; so that the number required is (as to its significant figures) within a unit of 51503. Let it be $51503 + h$, then we know that h is to be so taken* that

$$\log(51503 + h) = 4{\cdot}7118366$$

But (page 242),

$$\log(51503 + h) = \log 51503 + {\cdot}0000084 \times h \text{ very nearly.}$$
$$= 4{\cdot}7118325 + {\cdot}0000084 \times h$$

* We neglect the characteristic, or rather we make our logarithm 4·7118366. But an alteration of the characteristic is only an alteration of the place of the decimal point.

whence

$$h = \frac{4{\cdot}7118366 - 4{\cdot}7118325}{{\cdot}0000084} = \frac{{\cdot}0000041}{{\cdot}0000084} = \frac{41}{84}$$

Now, from the table of differences we see that

$$34 \text{ is } \frac{4}{10} \text{ of } 84 \text{ nearly,}$$

$$\begin{cases} 76 \text{ is } \frac{8}{10} \text{ of } 84, \text{ or} \\ 7 \text{ is } \frac{8}{100} \text{ of } 84 \text{ nearly;} \end{cases}$$

so that $34 + 7$ or 41 is $\frac{4}{10} + \frac{8}{100}$ of 84; that is, $\frac{41}{84} = {\cdot}48 = h$:
whence $51503 + h = 51503{\cdot}48$ and

4·7118366	is the log of	51503·84
$\bar{1}$·7118366	......	·5150384

But the readiest method of putting this into practice is by making an inverse process to the method of finding a logarithm. We take an instance from another part of the table.

What is the logarithm of 217483·6?

log 21748	(*)	5·3374193
3		60
6	(†)	12
log 217483·6	is	5·3374265

What is the number whose logarithm is 5·3374265?

	·3374265
Nearest log in table, belonging to N° 21748	·3374193
Difference	72
Nearest N° in table of Diff. belonging to N° 3	60
	120

Annex a cipher, because a figure was struck off, and therefore a figure (we do not know what) must be annexed. (See subsequent remark). Opposite to this we find 6.

* Observe that we put in the right characteristic at once.

† Here, as before, we look opposite to six, and find 120, from which we strike off one figure.

Therefore the number to the logarithm required is (making six places before the decimal point, on account of the characteristic 5) 217483·6.

The student will better understand, by forming a number of logarithms and inverting each process: 1. Why a figure must be annexed to the last difference, which comes after the table of differences has been used once. 2. Why that figure cannot be known. 3. Why 0 is most likely to be right. The above 12 might have been the result of any tabular difference between 115 and 125, the mean number of which is 120. The following are examples of the process, without explanation:

What are the numbers to the logarithms $\bar{2}\cdot1183214$ and $1\cdot9648317$?

	·1183214		·9648317
13131	·1182978	92221	·9648298
	236		19
7	232	4	19
	40		0
1	33		
Ans.	·01313171	Ans.	92·22140

The only unusual circumstance with which the student will now meet is in the multiplication and division of such quantities as $\bar{2}\cdot9$. To multiply this by 5, carry to the negative term according to the common algebraical rule of addition. Thus, five times 9 are 45, set down 5, and carry 4; five times -2 are -10, and 4 are -6. The answer then is $\bar{6}\cdot5$. Or

$$5(\cdot9-2) = 4\cdot5-10 = 4-10+\cdot5 = \bar{6}\cdot5$$

To divide $\bar{2}\cdot9$ by 5, make the negative term divisible by 5, and correct the expression by a corresponding addition. Thus, $\bar{2}\cdot9$ is $\bar{5}+3\cdot9$, and

$$\frac{\bar{2}\cdot9}{5} = \frac{\bar{5}}{5} + \frac{3\cdot9}{5} = \bar{1} + \cdot78 = \bar{1}\cdot78$$

The following are further instances:

$\bar{3}\cdot46$	$\bar{1}\cdot417$
8	10
$6)\overline{21}\cdot68$	$5)\bar{6}\cdot170$
$\bar{4}\cdot613$ ..	$\bar{2}\cdot834$

The following is an example of the use of logarithms in multiplication and division, &c. What is the result of ·5729578 multiplied by 20·62648, and the product divided by 7853982, after which the tenth root of the ninth power is taken? or what is

$$\left\{\frac{\cdot5729578 \times 20\cdot62648}{7853982}\right\}^{\frac{9}{10}}$$

log ·5729578		$\bar{1}\cdot7581226$
log 20·62648		$1\cdot3144251$
	Add	$1\cdot0725477$
log 7853982		$6\ 8950899$
	Subtract	$\bar{6}\cdot1774578$
		9
		$10)\overline{53}\cdot5971202$
		$\bar{6}\cdot7597120$
	57505	·7597056
		64
	8	60
		40
	5	38

Answer ·000005750585

The student should furnish himself with examples for practice by verifying such equations, for instance, as $a(a+b) = a^2 + ab$, where a and b may be any given numbers. The logarithms of a and $a+b$ being added together, and $a(a+b)$ thus found, a^2 and ab should be separately found; and if the whole be correct, the sum of the two last will be equal to the first.

The following equations may be thus used:

$$(a+b)(a-b) = a^2 - b^2$$

$$\sqrt{ab} = \sqrt{a} \times \sqrt{b}$$

$$(ab)^{\frac{m}{n}} = a^{\frac{m}{n}} \times b^{\frac{m}{n}}$$

Nothing but practice will enable the student to work correctly with logarithms, and most treatises on that subject contain detailed examples of all the cases which arise in practice.

THE END.

LONDON:
PRINTED BY JAMES MOYES, CASTLE STREET,
LEICESTER SQUARE.

ADDENDA ET CORRIGENDA.

Page viii, line 9, The mistake alluded to is the saying that a multiplied by *nothing* is a.

... x, ... 11, omit the word *always*.

... xii, ... 16, *for* that 1 contains, *read* that is, 1 contains.

... xx, ... 7, *for* shews, *read* shew.

... 3, ... 4 from the end, *for* added, *read* added, &c.

... 21, ... 8 from the end, *to* on each share, *add* which he holds.

... 24, ... 11, 12, and 14, *for* 1988 and 6988, *read* 1998 and 6998.

... 35, ... 4, *for* as, *read* as often as.

... 36, ... 20, *for* right, *read* left.

... 47, ... 9, *for* $(a+b)$, *read* $(a+c)$.

... 60, ... 6, *for* C, *read* A.

... 89, ... 13 from the end, *for* square root, *read* square.

... 103, Though the student omit what is said on the *law of continuity*, he should not omit either this page, or what immediately relates to it in the next.

... 109, ... 4, *for* $\sqrt{x^2}$, *read* $\sqrt[3]{x^4}$.

... 109, ... 10, *for* $x^{\frac{1}{6}}$, *read* $x^{\frac{2}{3}-\frac{1}{2}}$.

... 131, ... 7, *for* q, *read* r, in both places.

... 131, ... 9 and 11, *for* q, *read* r.

... 144, ... 7 from the end, *for* $b+\frac{1}{2v}$, *read* $b+\frac{v}{2b}$.

VI.

LESSONS ON FORM, as given at a Pestalozzian School at Cheam, Surrey. By CHARLES REINER, Author of "Lessons on Number." With numerous Diagrams. 12mo. 6*s*. cloth.

VII.

PRINCIPLES OF GEOMETRY, familiarly illustrated, and applied to a variety of useful purposes. Designed for the Instruction of Young Persons. By the late Rev. WILLIAM RITCHIE, LL.D., F.R.S. Second Edition, revised and enlarged. 12mo. with 150 Wood Cuts, 3*s*. 6*d*. cloth.

"The practical applications which are added, must render the study very delightful to the young, since the Exercises on the Principles will be found as amusing as the ordinary sports of childhood."—*Athenæum*, Sept. 28, 1833.

"Dr. Ritchie's little elementary work is excellently well adapted to its object. It is brief, plain, and full of all that is necessary; curious and useful in its application; and beyond any other of the kind now existent in its familiar and distinct explanation of some of the instruments required in the practical application of the principles laid down and demonstrated."—*Spectator*, Sept. 7, 1833.

AN INSTRUMENT FOR TEACHING GEOMETRY, convertible into a Theodolite, Spirit Level, Hadley's Sextant, and Wollaston's Goniometer, has been prepared, and may be had of the publishers. Price 2*l*. 12*s*. 6*d*.

VIII.

PRINCIPLES OF THE DIFFERENTIAL AND INTEGRAL CALCULUS, familiarly Illustrated, and applied to a variety of Useful Purposes. Designed for the Instruction of Young Persons. By the late Rev. WILLIAM RITCHIE, LL.D., F.R.S. With Wood Cuts. 12mo. 4*s*. 6*d*. cloth.

IX.

THE ELEMENTS OF EUCLID, with a Commentary and Geometrical Exercises. By the Rev. DIONYSIUS LARDNER, LL.D. Fifth Edition, 8vo. 7*s*. boards.

X.

AN ANALYTICAL TREATISE ON PLANE AND SPHERICAL TRIGONOMETRY. By the Rev. DIONYSIUS LARDNER, LL.D. Second Edition. Corrected and improved. 8vo. 12*s*. cloth.

XI.

AN ELEMENTARY TREATISE ON THE DIFFERENTIAL AND INTEGRAL CALCULUS. By the Rev. DIONYSIUS LARDNER, LL.D. 8vo. 21*s*.

XII.

DARLEY'S SCIENTIFIC LIBRARY.

1. A SYSTEM OF POPULAR GEOMETRY. Containing in a few Lessons so much of the Elements of Euclid as is necessary and sufficient for a right understanding of every Art and Science in its leading Truths and general Principles. By GEORGE DARLEY, A.B. Fourth Edition. 4*s*. 6*d*. cloth.

2. COMPANION TO THE POPULAR GEOMETRY. In which the Elements of Abstract Science are familiarised, illustrated, and rendered practically useful to the various purposes of Life, with numerous Cuts. 4*s*. 6*d*. cloth.

3. A SYSTEM OF POPULAR ALGEBRA, with a Section on Proportions and Progressions. Third Edition. 4*s*. 6*d*. cloth.

4. A SYSTEM OF POPULAR TRIGONOMETRY, both Plain and Spherical; with Popular Treatises on Logarithms, and the Application of Algebra to Geometry. Second Edition. 3*s*. 6*d*. cloth.

"For Students who only seek this limited knowledge of these Sciences, there are, perhaps, no Treatises which can be read with more advantage than Darley's Popular Geometry and Algebra."—*Library of Useful Knowledge, Article "Mechanics."*

Printed in Dunstable, United Kingdom

91218503R00171